Ground Water and Surface Water
A Single Resource

U.S. Geological Survey Circular 1139

by Thomas C. Winter
Judson W. Harvey
O. Lehn Franke
William M. Alley

Denver, Colorado
1998

U.S. DEPARTMENT OF THE INTERIOR
BRUCE BABBITT, Secretary

U.S. GEOLOGICAL SURVEY
Thomas J. Casadevall, Acting Director

The use of firm, trade, and brand names in this report is for identification purposes only and does not constitute endorsement by the U.S. Government

U.S. GOVERNMENT PRINTING OFFICE : 1998

Free on application to the
U.S. Geological Survey
Branch of Information Services
Box 25286
Denver, CO 80225-0286

Library of Congress Cataloging-in-Publications Data

Ground water and surface water : a single resource /
  by Thomas C. Winter ... [et al.].
    p.   cm. -- (U.S. Geological Survey circular : 1139)
  Includes bibliographical references.
  1. Hydrology. I. Winter, Thomas C. II. Series.
GB661.2.G76 1998   98–2686
553.7—dc21   CIP
                                ISBN 0–607–89339–7

FOREWORD

Traditionally, management of water resources has focused on surface water or ground water as if they were separate entities. As development of land and water resources increases, it is apparent that development of either of these resources affects the quantity and quality of the other. Nearly all surface-water features (streams, lakes, reservoirs, wetlands, and estuaries) interact with ground water. These interactions take many forms. In many situations, surface-water bodies gain water and solutes from ground-water systems and in others the surface-water body is a source of ground-water recharge and causes changes in ground-water quality. As a result, withdrawal of water from streams can deplete ground water or conversely, pumpage of ground water can deplete water in streams, lakes, or wetlands. Pollution of surface water can cause degradation of ground-water quality and conversely pollution of ground water can degrade surface water. Thus, effective land and water management requires a clear understanding of the linkages between ground water and surface water as it applies to any given hydrologic setting.

This Circular presents an overview of current understanding of the interaction of ground water and surface water, in terms of both quantity and quality, as applied to a variety of landscapes across the Nation. This Circular is a product of the Ground-Water Resources Program of the U.S. Geological Survey. It serves as a general educational document rather than a report of new scientific findings. Its intent is to help other Federal, State, and local agencies build a firm scientific foundation for policies governing the management and protection of aquifers and watersheds. Effective policies and management practices must be built on a foundation that recognizes that surface water and ground water are simply two manifestations of a single integrated resource. It is our hope that this Circular will contribute to the use of such effective policies and management practices.

Robert M. Hirsch
Chief Hydrologist

CONTENTS

Preface   VI
Introduction   1
Natural processes of ground-water and surface-water interaction   2
    The hydrologic cycle and interactions of ground water and surface water   2
        Interaction of ground water and streams   9
        Interaction of ground water and lakes   18
        Interaction of ground water and wetlands   19
    Chemical interactions of ground water and surface water   22
        Evolution of water chemistry in drainage basins   22
        Chemical interactions of ground water and surface water in streams, lakes, and wetlands   23
    Interaction of ground water and surface water in different landscapes   33
        Mountainous terrain   33
        Riverine terrain   38
        Coastal terrain   42
        Glacial and dune terrain   46
        Karst terrain   50
Effects of human activities on the interaction of ground water and surface water   54
    Agricultural development   54
        Irrigation systems   57
        Use of agricultural chemicals   61
    Urban and industrial development   66
    Drainage of the land surface   67
    Modifications to river valleys   68
        Construction of levees   68
        Construction of reservoirs   68
        Removal of natural vegetation   69
    Modifications to the atmosphere   72
        Atmospheric deposition   72
        Global warming   72
Challenges and opportunities   76
    Water supply   76
    Water quality   77
    Characteristics of aquatic environments   78
Acknowledgments   79

BOXES

Box A -- Concepts of ground water, water table, and flow systems   6

Box B -- The ground-water component of streamflow   12

Box C -- The effect of ground-water withdrawals on surface water   14

Box D -- Some common types of biogeochemical reactions affecting transport of chemicals in ground water and surface water   24

Box E -- Evolution of ground-water chemistry from recharge to discharge areas in the Atlantic Coastal Plain   26

Box F -- The interface between ground water and surface water as an environmental entity   28

Box G -- Use of environmental tracers to determine the interaction of ground water and surface water   30

Box H -- Field studies of mountainous terrain   36

Box I -- Field studies of riverine terrain   40

Box J -- Field studies of coastal terrain   44

Box K -- Field studies of glacial and dune terrain   48

Box L -- Field studies of karst terrain   52

Box M -- Point and nonpoint sources of contaminants   56

Box N -- Effects of irrigation development on the interaction of ground water and surface water   58

Box O -- Effects of nitrogen use on the quality of ground water and surface water   62

Box P -- Effects of pesticide application to agricultural lands on the quality of ground water and surface water   64

Box Q -- Effects of surface-water reservoirs on the interaction of ground water and surface water   70

Box R -- Effects of the removal of flood-plain vegetation on the interaction of ground water and surface water   71

Box S -- Effects of atmospheric deposition on the quality of ground water and surface water   74

PREFACE

- Understanding the interaction of ground water and surface water is essential to water managers and water scientists. Management of one component of the hydrologic system, such as a stream or an aquifer, commonly is only partly effective because each hydrologic component is in continuing interaction with other components. The following are a few examples of common water-resource issues where understanding the interconnections of ground water and surface water is fundamental to development of effective water-resource management and policy.

WATER SUPPLY

- It has become difficult in recent years to construct reservoirs for surface storage of water because of environmental concerns and because of the difficulty in locating suitable sites. An alternative, which can reduce or eliminate the necessity for surface storage, is to use an aquifer system for temporary storage of water. For example, water stored underground during times of high streamflow can be withdrawn during times of low streamflow. The characteristics and extent of the interactions of ground water and surface water affect the success of such conjunctive-use projects.

- Methods of accounting for water rights of streams invariably account for surface-water diversions and surface-water return flows. Increasingly, the diversions from a stream that result from ground-water withdrawals are considered in accounting for water rights as are ground-water return flows from irrigation and other applications of water to the land surface. Accounting for these ground-water components can be difficult and controversial. Another form of water-rights accounting involves the trading of ground-water rights and surface-water rights. This has been proposed as a water-management tool where the rights to the total water resource can be shared. It is an example of the growing realization that ground water and surface water are essentially one resource.

- In some regions, the water released from reservoirs decreases in volume, or is delayed significantly, as it moves downstream because some of the released water seeps into the streambanks. These losses of water and delays in traveltime can be significant, depending on antecedent ground-water and streamflow conditions as well as on other factors such as the condition of the channel and the presence of aquatic and riparian vegetation.

- Storage of water in streambanks, on flood plains, and in wetlands along streams reduces flooding downstream. Modifications of the natural interaction between ground water and surface water along streams, such as drainage of wetlands and construction of levees, can remove some of this natural attenuation of floods. Unfortunately, present knowledge is limited with respect to the effects of land-surface modifications in river valleys on floods and on the natural interaction of ground water and surface water in reducing potential flooding.

WATER QUALITY

- Much of the ground-water contamination in the United States is in shallow aquifers that are directly connected to surface water. In some settings where this is the case, ground water can be a major and potentially long-term contributor to contamination of surface water. Determining the contributions of ground water to contamination of streams and lakes is a critical step in developing effective water-management practices.

- A focus on watershed planning and management is increasing among government agencies responsible for managing water quality as well as broader aspects of the environment. The watershed approach recognizes that water, starting with precipitation, usually moves through the subsurface before entering stream channels and flowing out of the watershed. Integrating ground water into this "systems" approach is essential, but challenging, because of limitations in knowledge of the interactions of ground water and surface water. These difficulties are further complicated by the fact that surface-water watersheds and ground-water watersheds may not coincide.

- To meet water-quality standards and criteria, States and local agencies need to determine the amount of contaminant movement (wasteload) to surface waters so they can issue permits and control discharges of waste. Typically, ground-water inputs are not included in estimates of wasteload; yet, in some cases, water-quality standards and criteria cannot be met without reducing contaminant loads from ground-water discharges to streams.

- It is generally assumed that ground water is safe for consumption without treatment. Concerns about the quality of ground water from wells near streams, where contaminated surface water might be part of the source of water to the well, have led to increasing interest in identifying when filtration or treatment of ground water is needed.

- Wetlands, marshes, and wooded areas along streams (riparian zones) are protected in some areas to help maintain wildlife habitat and the quality of nearby surface water. Greater knowledge of the water-quality functions of riparian zones and of the pathways of exchange between shallow ground water and surface-water bodies is necessary to properly evaluate the effects of riparian zones on water quality.

CHARACTERISTICS OF AQUATIC ENVIRONMENTS

- Mixing of ground water with surface water can have major effects on aquatic environments if factors such as acidity, temperature, and dissolved oxygen are altered. Thus, changes in the natural interaction of ground water and surface water caused by human activities can potentially have a significant effect on aquatic environments.

- The flow between surface water and ground water creates a dynamic habitat for aquatic fauna near the interface. These organisms are part of a food chain that sustains a diverse ecological community. Studies indicate that these organisms may provide important indications of water quality as well as of adverse changes in aquatic environments.

- Many wetlands are dependent on a relatively stable influx of ground water throughout changing seasonal and annual weather patterns. Wetlands can be highly sensitive to the effects of ground-water development and to land-use changes that modify the ground-water flow regime of a wetland area. Understanding wetlands in the context of their associated ground-water flow systems is essential to assessing the cumulative effects of wetlands on water quality, ground-water flow, and streamflow in large areas.

- The success of efforts to construct new wetlands that replicate those that have been destroyed depends on the extent to which the replacement wetland is hydrologically similar to the destroyed wetland. For example, the replacement of a wetland that is dependent on ground water for its water and chemical input needs to be located in a similar ground-water discharge area if the new wetland is to replicate the original. Although a replacement wetland may have a water depth similar to the original, the communities that populate the replacement wetland may be completely different from communities that were present in the original wetland because of differences in hydrogeologic setting.

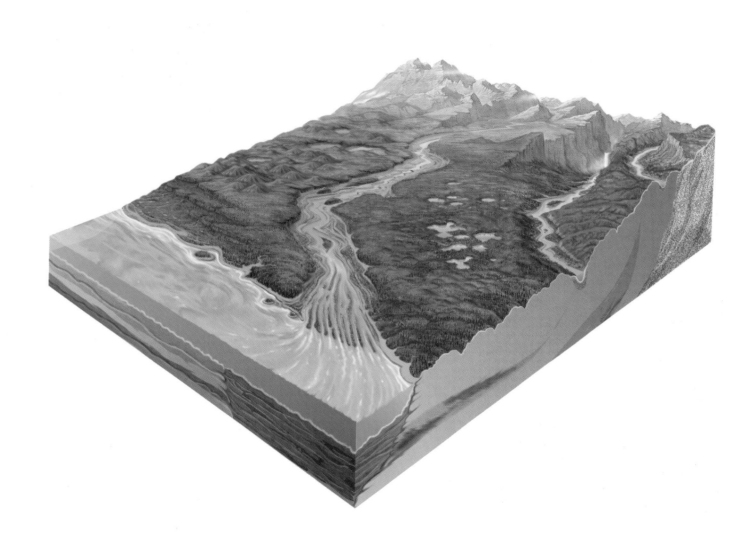

Ground Water and Surface Water
A Single Resource

by T.C. Winter
J.W. Harvey
O.L. Franke
W.M. Alley

INTRODUCTION

As the Nation's concerns over water resources and the environment increase, the importance of considering ground water and surface water as a single resource has become increasingly evident. Issues related to water supply, water quality, and degradation of aquatic environments are reported on frequently. The interaction of ground water and surface water has been shown to be a significant concern in many of these issues. For example, contaminated aquifers that discharge to streams can result in long-term contamination of surface water; conversely, streams can be a major source of contamination to aquifers. Surface water commonly is hydraulically connected to ground water, but the interactions are difficult to observe and measure and commonly have been ignored in water-management considerations and policies. Many natural processes and human activities affect the interactions of ground water and surface water. The purpose of this report is to present our current understanding of these processes and activities as well as limitations in our knowledge and ability to characterize them.

"Surface water commonly is hydraulically connected to ground water, but the interactions are difficult to observe and measure"

NATURAL PROCESSES OF GROUND-WATER AND SURFACE-WATER INTERACTION

The Hydrologic Cycle and Interactions of Ground Water and Surface Water

The hydrologic cycle describes the continuous movement of water above, on, and below the surface of the Earth. The water on the Earth's surface—surface water—occurs as streams, lakes, and wetlands, as well as bays and oceans. Surface water also includes the solid forms of water—snow and ice. The water below the surface of the Earth primarily is ground water, but it also includes soil water.

The hydrologic cycle commonly is portrayed by a very simplified diagram that shows only major transfers of water between continents and oceans, as in Figure 1. However, for understanding hydrologic processes and managing water resources, the hydrologic cycle needs to be viewed at a wide range of scales and as having a great deal of variability in time and space. Precipitation, which is the source of virtually all freshwater in the hydrologic cycle, falls nearly everywhere, but its distribution is highly variable. Similarly, evaporation and transpiration return water to the atmosphere nearly everywhere, but evaporation and transpiration rates vary considerably according to climatic conditions. As a result, much of the precipitation never reaches the oceans as surface and subsurface runoff before the water is returned to the atmosphere. The relative magnitudes of the individual components of the hydrologic cycle, such as evapotranspiration, may differ significantly even at small scales, as between an agricultural field and a nearby woodland.

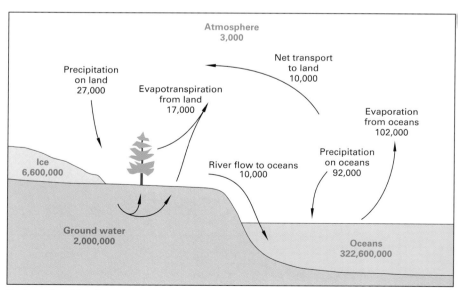

Figure 1. Ground water is the second smallest of the four main pools of water on Earth, and river flow to the oceans is one of the smallest fluxes, yet ground water and surface water are the components of the hydrologic system that humans use most. (Modified from Schelesinger, W.H., 1991, Biogeochemistry–An analysis of global change: Academic Press, San Diego, California.) (Used with permission.)

To present the concepts and many facets of the interaction of ground water and surface water in a unified way, a conceptual landscape is used (Figure 2). The conceptual landscape shows in a very general and simplified way the interaction of ground water with all types of surface water, such as streams, lakes, and wetlands, in many different terrains from the mountains to the oceans. The intent of Figure 2 is to emphasize that ground water and surface water interact at many places throughout the landscape.

Movement of water in the atmosphere and on the land surface is relatively easy to visualize, but the movement of ground water is not. Concepts related to ground water and the movement of ground water are introduced in Box A. As illustrated in Figure 3, ground water moves along flow paths of varying lengths from areas of recharge to areas of discharge. The generalized flow paths in Figure 3 start at the water table, continue through the ground-water system, and terminate at the stream or at the pumped well. The source of water to the water table (ground-water recharge) is infiltration of precipitation through the unsaturated zone. In the uppermost, unconfined aquifer, flow paths near the stream can be tens to hundreds of feet in length and have corresponding traveltimes of days to a few years. The longest and deepest flow paths in Figure 3 may be thousands of feet to tens of miles in length, and traveltimes may range from decades to millennia. In general, shallow ground water is more susceptible to contamination from human sources and activities because of its close proximity to the land surface. Therefore, shallow, local patterns of ground-water flow near surface water are emphasized in this Circular.

Haze over Appalachian Mountains in North Carolina. (Photograph courtesy of North Carolina Department of Travel and Tourism.)

"Ground water moves along flow paths of varying lengths in transmitting water from areas of recharge to areas of discharge"

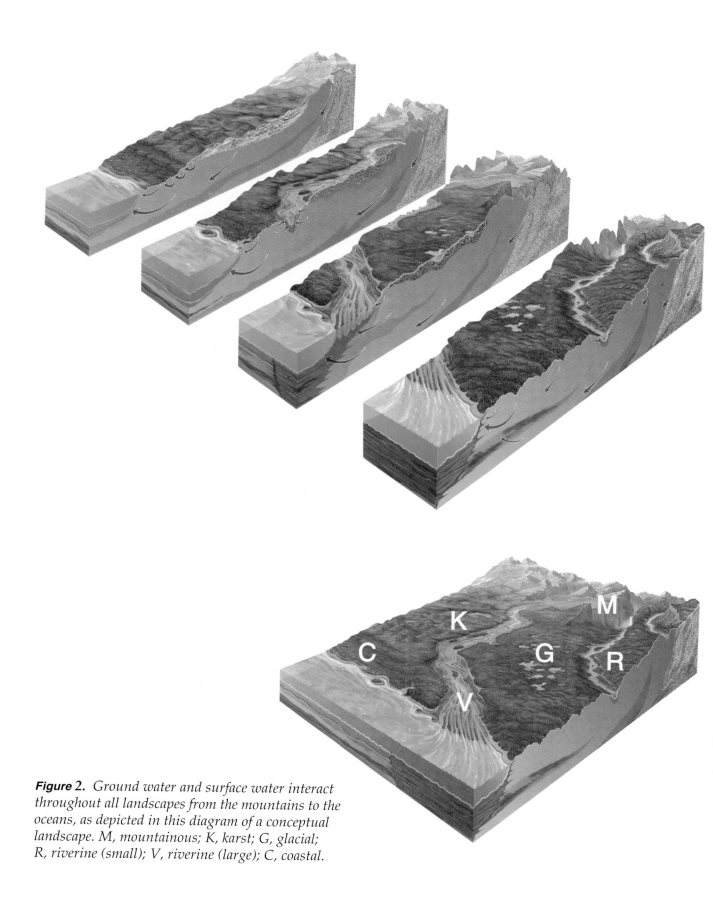

Figure 2. Ground water and surface water interact throughout all landscapes from the mountains to the oceans, as depicted in this diagram of a conceptual landscape. M, mountainous; K, karst; G, glacial; R, riverine (small); V, riverine (large); C, coastal.

Small-scale geologic features in beds of surface-water bodies affect seepage patterns at scales too small to be shown in Figure 3. For example, the size, shape, and orientation of the sediment grains in surface-water beds affect seepage patterns. If a surface-water bed consists of one sediment type, such as sand, inflow seepage is greatest at the shoreline, and it decreases in a nonlinear pattern away from the shoreline (Figure 4). Geologic units having different permeabilities also affect seepage distribution in surface-water beds. For example, a highly permeable sand layer within a surface-water bed consisting largely of silt will transmit water preferentially into the surface water as a spring (Figure 5).

Subaqueous spring in Nebraska. (Photograph by Charles Flowerday.)

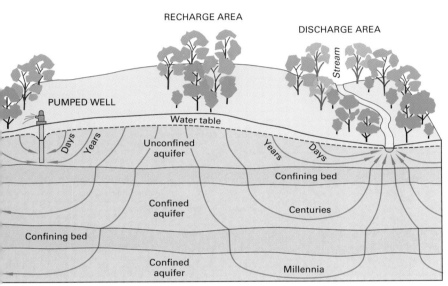

Figure 3. Ground-water flow paths vary greatly in length, depth, and traveltime from points of recharge to points of discharge in the ground-water system.

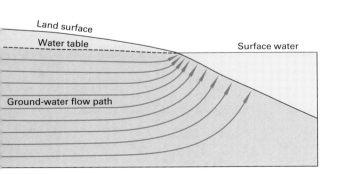

Figure 4. Ground-water seepage into surface water usually is greatest near shore. In flow diagrams such as that shown here, the quantity of discharge is equal between any two flow lines; therefore, the closer flow lines indicate greater discharge per unit of bottom area.

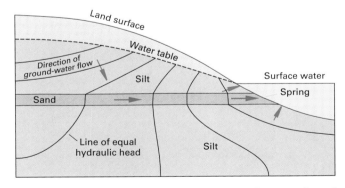

Figure 5. Subaqueous springs can result from preferred paths of ground-water flow through highly permeable sediments.

Concepts of Ground Water, Water Table, and Flow Systems

SUBSURFACE WATER

Water beneath the land surface occurs in two principal zones, the unsaturated zone and the saturated zone (Figure A–1). In the unsaturated zone, the voids—that is, the spaces between grains of gravel, sand, silt, clay, and cracks within rocks—contain both air and water. Although a considerable amount of water can be present in the unsaturated zone, this water cannot be pumped by wells because it is held too tightly by capillary forces. The upper part of the unsaturated zone is the soil-water zone. The soil zone is crisscrossed by roots, voids left by decayed roots, and animal and worm burrows, which enhance the infiltration of precipitation into the soil zone. Soil water is used by plants in life functions and transpiration, but it also can evaporate directly to the atmosphere.

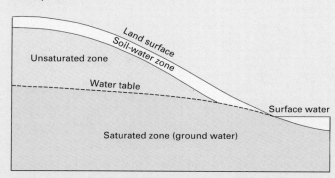

Figure A–1. *The water table is the upper surface of the saturated zone. The water table meets surface-water bodies at or near the shoreline of surface water if the surface-water body is connected to the ground-water system.*

In contrast to the unsaturated zone, the voids in the saturated zone are completely filled with water. Water in the saturated zone is referred to as ground water. The upper surface of the saturated zone is referred to as the water table. Below the water table, the water pressure is great enough to allow water to enter wells, thus permitting ground water to be withdrawn for use. A well is constructed by inserting a pipe into a drilled hole; a screen is attached, generally at its base, to prevent earth materials from entering the pipe along with the water pumped through the screen.

The depth to the water table is highly variable and can range from zero, when it is at land surface, to hundreds or even thousands of feet in some types of landscapes. Usually, the depth to the water table is small near permanent bodies of surface water such as streams, lakes, and wetlands. An important characteristic of the water table is that its configuration varies seasonally and from year to year because ground-water recharge, which is the accretion of water to the upper surface of the saturated zone, is related to the wide variation in the quantity, distribution, and timing of precipitation.

THE WATER TABLE

The depth to the water table can be determined by installing wells that penetrate the top of the saturated zone just far enough to hold standing water. Preparation of a water table map requires that only wells that have their well screen placed near the water table be used. If the depth to water is measured at a number of such wells throughout an area of study, and if those water levels are referenced to a common datum such as sea level, the data can be contoured to indicate the configuration of the water table (Figure A–2).

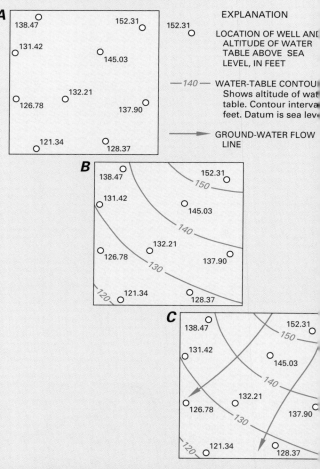

Figure A–2. *Using known altitudes of the water table at individual wells (A), contour maps of the water-table surface can be drawn (B), and directions of ground-water flow along the water table can be determined (C) because flow usually is approximately perpendicular to the contours.*

In addition to various practical uses of a water-table map, such as estimating an approximate depth for a proposed well, the configuration of the water table provides an indication of the approximate direction of ground-water flow at any location on the water table. Lines drawn perpendicular to water-table contours usually indicate the direction of ground-water flow along the upper surface of the ground-water system. The water table is continually adjusting to changing recharge and discharge patterns. Therefore, to construct a water-table map, water-level measurements must be made at approximately the same time, and the resulting map is representative only of that specific time.

GROUND-WATER MOVEMENT

The ground-water system as a whole is actually a three-dimensional flow field; therefore, it is important to understand how the vertical components of ground-water movement affect the interaction of ground water and surface water. A vertical section of a flow field indicates how potential energy is distributed beneath the water table in the ground-water system and how the energy distribution can be used to determine vertical components of flow near a surface-water body. The term hydraulic head, which is the sum of elevation and water pressure divided by the weight density of water, is used to describe potential energy in ground-water flow systems. For example, Figure A–3 shows a generalized vertical section of subsurface water flow. Water that infiltrates at land surface moves vertically downward to the water table to become ground water. The ground water then moves both vertically and laterally within the ground-water system. Movement is downward and lateral on the right side of the diagram, mostly lateral in the center, and lateral and upward on the left side of the diagram.

Flow fields such as these can be mapped in a process similar to preparing water-table maps, except that vertically distributed piezometers need to be used instead of water-table wells. A piezometer is a well that has a very short screen so the water level represents hydraulic head in only a very small part of the ground-water system. A group of piezometers completed at different depths at the same location is referred to as a piezometer nest. Three such piezometer nests are shown in Figure A–3 (locations A, B, and C). By starting at a water-table contour, and using the water-level data from the piezometer nests, lines of equal hydraulic head can be drawn. Similar to drawing flow direction on water-table maps, flow lines can be drawn approximately perpendicular to these lines of equal hydraulic head, as shown in Figure A–3.

Actual flow fields generally are much more complex than that shown in Figure A–3. For example, flow systems of different sizes and depths can be present, and they can overlie one another, as indicated in Figure A–4. In a local flow system, water that recharges at a water-table high discharges to an adjacent lowland. Local flow systems are the most dynamic and the shallowest flow systems; therefore, they have the greatest interchange with surface water. Local flow systems can be underlain by intermediate and regional flow systems. Water in deeper flow systems have longer flow paths and longer contact time with subsurface materials; therefore, the water generally contains more dissolved chemicals. Nevertheless, these deeper flow systems also eventually discharge to surface water, and they can have a great effect on the chemical characteristics of the receiving surface water.

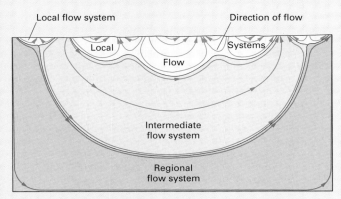

Figure A–4. *Ground-water flow systems can be local, intermediate, and regional in scale. Much ground-water discharge into surface-water bodies is from local flow systems. (Figure modified from Toth, J., 1963, A theoretical analysis of groundwater flow in small drainage basins: p. 75–96 in Proceedings of Hydrology Symposium No. 3, Groundwater, Queen's Printer, Ottawa, Canada.)*

GROUND-WATER DISCHARGE

The quantity of ground-water discharge (flux) to and from surface-water bodies can be determined for a known cross section of aquifer by multiplying the hydraulic gradient, which is determined from the hydraulic-head measurements in wells and piezometers, by the permeability of the aquifer materials. Permeability is a quantitative measure of the ease of water movement through aquifer materials. For example, sand is more permeable than clay because the pore spaces between sand grains are larger than pore spaces between clay particles.

Figure A–3. *If the distribution of hydraulic head in vertical section is known from nested piezometer data, zones of downward, lateral, and upward components of ground-water flow can be determined.*

Changing meteorological conditions also strongly affect seepage patterns in surface-water beds, especially near the shoreline. The water table commonly intersects land surface at the shoreline, resulting in no unsaturated zone at this point. Infiltrating precipitation passes rapidly through a thin unsaturated zone adjacent to the shoreline, which causes water-table mounds to form quickly adjacent to the surface water (Figure 6). This process, termed focused recharge, can result in increased ground-water inflow to surface-water bodies, or it can cause inflow to surface-water bodies that normally have seepage to ground water. Each precipitation event has the potential to cause this highly transient flow condition near shorelines as well as at depressions in uplands (Figure 6).

These periodic changes in the direction of flow also take place on longer time scales: focused recharge from precipitation predominates during wet periods and drawdown by transpiration predominates during dry periods. As a result, the two processes, together with the geologic controls on seepage distribution, can cause flow conditions at the edges of surface-water bodies to be extremely variable. These "edge effects" probably affect small surface-water bodies more than large surface-water bodies because the ratio of edge length to total volume is greater for small water bodies than it is for large ones.

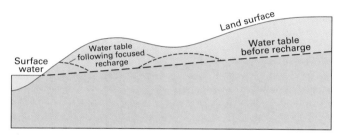

Figure 6. *Ground-water recharge commonly is focused initially where the unsaturated zone is relatively thin at the edges of surface-water bodies and beneath depressions in the land surface.*

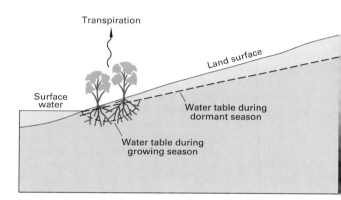

Figure 7. *Where the depth to the water table is small adjacent to surface-water bodies, transpiration directly from ground water can cause cones of depression similar to those caused by pumping wells. This sometimes draws water directly from the surface water into the subsurface.*

Transpiration by nearshore plants has the opposite effect of focused recharge. Again, because the water table is near land surface at edges of surface-water bodies, plant roots can penetrate into the saturated zone, allowing the plants to transpire water directly from the ground-water system (Figure 7). Transpiration of ground water commonly results in a drawdown of the water table much like the effect of a pumped well. This highly variable daily and seasonal transpiration of ground water may significantly reduce ground-water discharge to a surface-water body or even cause movement of surface water into the subsurface. In many places it is possible to measure diurnal changes in the direction of flow during seasons of active plant growth; that is, ground water moves into the surface water during the night, and surface water moves into shallow ground water during the day.

Phreatophytes along the Rio Grande in Texas. (Photograph by Michael Collier.)

INTERACTION OF GROUND WATER AND STREAMS

Streams interact with ground water in all types of landscapes (see Box B). The interaction takes place in three basic ways: streams gain water from inflow of ground water through the streambed (gaining stream, Figure 8A), they lose water to ground water by outflow through the streambed (losing stream, Figure 9A), or they do both, gaining in some reaches and losing in other reaches. For ground water to discharge into a stream channel, the altitude of the water table in the vicinity of the stream must be higher than the altitude of the stream-water surface. Conversely, for surface water to seep to ground water, the altitude of the water table in the vicinity of the stream must be lower than the altitude of the stream-water surface. Contours of water-table elevation indicate gaining streams by pointing in an upstream direction (Figure 8B), and they indicate losing streams by pointing in a downstream direction (Figure 9B) in the immediate vicinity of the stream.

Losing streams can be connected to the ground-water system by a continuous saturated zone (Figure 9A) or can be disconnected from

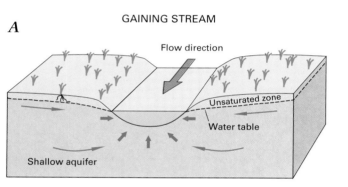

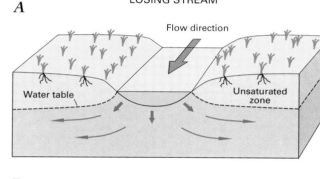

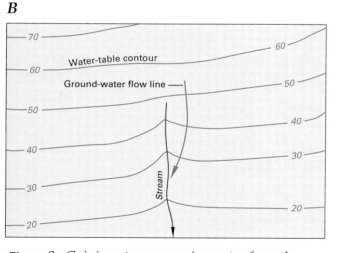

Figure 8. *Gaining streams receive water from the ground-water system (A). This can be determined from water-table contour maps because the contour lines point in the upstream direction where they cross the stream (B).*

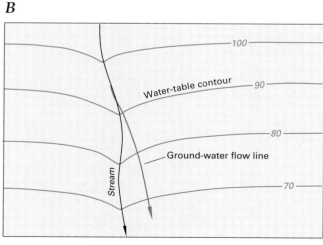

Figure 9. *Losing streams lose water to the ground-water system (A). This can be determined from water-table contour maps because the contour lines point in the downstream direction where they cross the stream (B).*

the ground-water system by an unsaturated zone. Where the stream is disconnected from the ground-water system by an unsaturated zone, the water table may have a discernible mound below the stream (Figure 10) if the rate of recharge through the streambed and unsaturated zone is greater than the rate of lateral ground-water flow away from the water-table mound. An important feature of streams that are disconnected from ground water is that pumping of shallow ground water near the stream does not affect the flow of the stream near the pumped wells.

In some environments, streamflow gain or loss can persist; that is, a stream might always gain water from ground water, or it might always lose water to ground water. However, in other environments, flow direction can vary a great deal along a stream; some reaches receive ground water, and other reaches lose water to ground water. Furthermore, flow direction can change in very short timeframes as a result of individual storms causing focused recharge near the streambank, temporary flood peaks moving down the channel, or transpiration of ground water by streamside vegetation.

A type of interaction between ground water and streams that takes place in nearly all streams at one time or another is a rapid rise in stream stage that causes water to move from the stream into the streambanks. This process, termed bank storage (Figures 11 and 12*B*), usually is caused by storm precipitation, rapid snowmelt, or release of

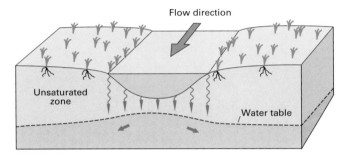

Figure 10. *Disconnected streams are separated from the ground-water system by an unsaturated zone.*

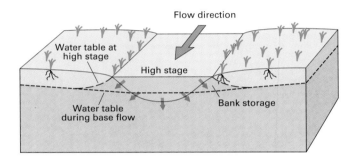

Figure 11. *If stream levels rise higher than adjacent ground-water levels, stream water moves into the streambanks as bank storage.*

"Streams interact with ground water in three basic ways: streams gain water from inflow of ground water through the streambed (gaining stream), they lose water to ground water by outflow through the streambed (losing stream), or they do both, gaining in some reaches and losing in other reaches"

10

water from a reservoir upstream. As long as the rise in stage does not overtop the streambanks, most of the volume of stream water that enters the streambanks returns to the stream within a few days or weeks. The loss of stream water to bank storage and return of this water to the stream in a period of days or weeks tends to reduce flood peaks and later supplement stream flows. If the rise in stream stage is sufficient to overtop the banks and flood large areas of the land surface, widespread recharge to the water table can take place throughout the flooded area (Figure 12C). In this case, the time it takes for the recharged floodwater to return to the stream by ground-water flow may be weeks, months, or years because the lengths of the ground-water flow paths are much longer than those resulting from local bank storage. Depending on the frequency, magnitude, and intensity of storms and on the related magnitude of increases in stream stage, some streams and adjacent shallow aquifers may be in a continuous readjustment from interactions related to bank storage and overbank flooding.

In addition to bank storage, other processes may affect the local exchange of water between streams and adjacent shallow aquifers. Changes in streamflow between gaining and losing conditions can also be caused by pumping ground water near streams (see Box C). Pumping can

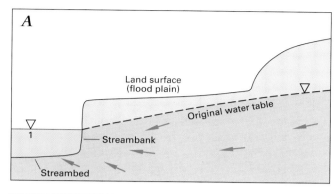

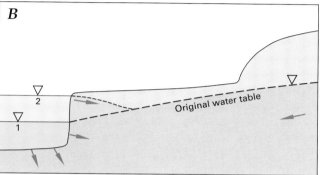

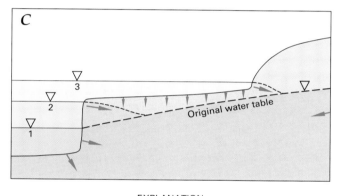

EXPLANATION

▽ ▽ ▽ Sequential stream stages
1 2 3

→ Approximate direction of ground-water flow or recharge through the unsaturated zone

Figure 12. *If stream levels rise higher than their streambanks (C), the floodwaters recharge ground water throughout the flooded areas.*

Flooding at the confluence of the Missouri and Mississippi Rivers. (Photograph by Robert Meade.)

intercept ground water that would otherwise have discharged to a gaining stream, or at higher pumping rates it can induce flow from the stream to the aquifer.

The Ground-Water Component of Streamflow

Ground water contributes to streams in most physiographic and climatic settings. Even in settings where streams are primarily losing water to ground water, certain reaches may receive ground-water inflow during some seasons. The proportion of stream water that is derived from ground-water inflow varies across physiographic and climatic settings. The amount of water that ground water contributes to streams can be estimated by analyzing streamflow hydrographs to determine the ground-water component, which is termed base flow (Figure B–1). Several different methods of analyzing hydrographs have been used by hydrologists to determine the base-flow component of streamflow.

One of the methods, which provides a conservative estimate of base flow, was used to determine the ground-water contribution to streamflow in 24 regions in the conterminous United States. The regions, delineated on the basis of physiography and climate, are believed to have common characteristics with respect to the interactions of ground water and surface water (Figure B–2). Fifty-four streams were selected for the analysis, at least two in each of the 24 regions. Streams were selected that had drainage basins less than 250 square miles and that had less than 3 percent of the drainage area covered by lakes and wetlands. Daily streamflow values for the 30-year period, 1961–1990, were used for the analysis of each stream. The analysis indicated that, for the 54 streams over the 30-year period, an average of 52 percent of the streamflow was contributed by ground water. Ground-water contributions ranged from 14 percent to 90 percent, and the median was 55 percent. The ground-water contribution to streamflow for selected streams can be compared in Figure B–2. As an example of the effect that geologic setting has on the contribution of ground water to streamflow, the Forest River in North Dakota can be compared to the Sturgeon River in Michigan. The Forest River Basin is underlain by poorly permeable silt and clay deposits, and only about 14 percent of its average annual flow is contributed by ground water; in contrast, the Sturgeon River Basin is underlain by highly permeable sand and gravel, and about 90 percent of its average annual flow is contributed by ground water.

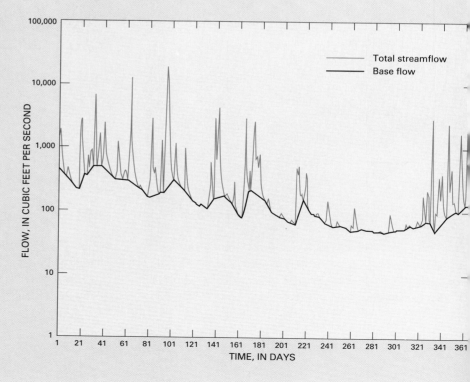

Figure B–1. The ground-water component of streamflow was estimated from a streamflow hydrograph for the Homochitto River in Mississippi, using a method developed by the institute of Hydrology, United Kingdom. (Institute of Hydrology, 1980, Low flow studies: Wallingford, Oxon, United Kingdom, Research Report No. 1.)

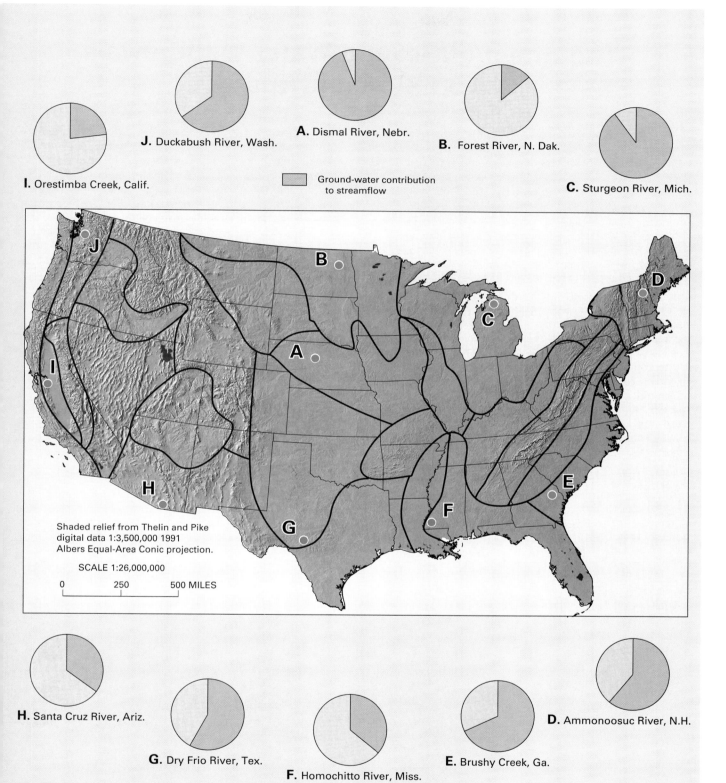

Figure B–2. In the conterminous United States, 24 regions were delineated where the interactions of ground water and surface water are considered to have similar characteristics. The estimated ground-water contribution to streamflow is shown for specific streams in 10 of the regions.

The Effect of Ground-Water Withdrawals on Surface Water

Withdrawing water from shallow aquifers that are directly connected to surface-water bodies can have a significant effect on the movement of water between these two water bodies. The effects of pumping a single well or a small group of wells on the hydrologic regime are local in scale. However, the effects of many wells withdrawing water from an aquifer over large areas may be regional in scale.

Withdrawing water from shallow aquifers for public and domestic water supply, irrigation, and industrial uses is widespread. Withdrawing water from shallow aquifers near surface-water bodies can diminish the available surface-water supply by capturing some of the ground-water flow that otherwise would have discharged to surface water or by inducing flow from surface water into the surrounding aquifer system. An analysis of the sources of water to a pumping well in a shallow aquifer that discharges to a stream is provided here to gain insight into how a pumping well can change the quantity and direction of flow between the shallow aquifer and the stream. Furthermore, changes in the direction of flow between the two water bodies can affect transport of contaminants associated with the moving water. Although a stream is used in the example, the results apply to all surface-water bodies, including lakes and wetlands.

A ground-water system under predevelopment conditions is in a state of dynamic equilibrium—for example, recharge at the water table is equal to ground-water discharge to a stream (Figure C–1A). Assume a well is installed and is pumped continually at a rate, Q_1. After a new state of dynamic equilibrium is achieved, inflow to the ground-water system from recharge will equal outflow to the stream plus the withdrawal from the well. In this new equilibrium, some of the ground water that would have discharged to the stream is intercepted by the well, and a ground-water divide, which is a line separating directions of flow, is established locally between the well and the stream (Figure C–1B). If the well is pumped at a higher rate, Q_2, at a later time a new equilibrium is reached. Under this condition, the ground-water divide between the well and the stream is no longer present and withdrawals from the well induce movement of water from the stream into the aquifer (Figure C–1C). Thus, pumpage reverses the hydrologic condition of the stream in this reach from a ground-water discharge feature to a ground-water recharge feature.

In the hydrologic system depicted in Figures C–1A and C–1B, the quality of the stream water generally will have little effect on the quality of the shallow ground water. However, in the case of the well pumping at the higher rate, Q_2 (Figure C–1C), the quality of the stream water, which locally recharges the shallow aquifer, can affect the quality of ground water between the well and the stream as well as the quality of the ground water withdrawn from the well.

This hypothetical withdrawal of water from a shallow aquifer that discharges to a nearby surface-water body is a simplified but compelling illustration of the concept that ground water and surface water are one resource. In the long term, the quantity of ground water withdrawn is approximately equal to the reduction in streamflow that is potentially available to downstream users.

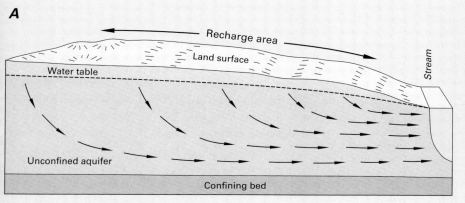
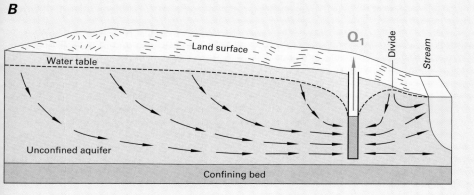
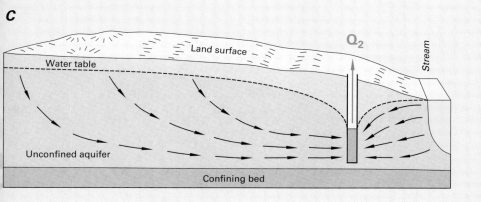

Figure C–1. In a schematic hydrologic setting where ground water discharges to a stream under natural conditions (A), placement of a well pumping at a rate (Q_1) near the stream will intercept part of the ground water that would have discharged to the stream (B). If the well is pumped at an even greater rate (Q_2), it can intercept additional water that would have discharged to the stream in the vicinity of the well and can draw water from the stream to the well (C).

Where streamflow is generated in headwaters areas, the changes in streamflow between gaining and losing conditions may be particularly variable (Figure 13). The headwaters segment of streams can be completely dry except during storm events or during certain seasons of the year when snowmelt or precipitation is sufficient to maintain continuous flow for days or weeks. During these times, the stream will lose water to the unsaturated zone beneath its bed. However, as the water table rises through recharge in the headwaters area, the losing reach may become a gaining reach as the water table rises above the level of the stream. Under these conditions, the point where ground water first contributes to the stream gradually moves upstream.

Some gaining streams have reaches that lose water to the aquifer under normal conditions of streamflow. The direction of seepage through the bed of these streams commonly is related to abrupt changes in the slope of the streambed (Figure 14A) or to meanders in the stream channel (Figure 14B). For example, a losing stream reach usually is located at the downstream end of pools in pool and riffle streams (Figure 14A), or upstream from channel bends in meandering streams (Figure 14B). The subsurface zone where stream water flows through short segments of its adjacent bed and banks is referred to as the hyporheic zone. The size and geometry of hyporheic zones surrounding streams vary greatly in time and space. Because of mixing between ground water and surface water in the hyporheic zone, the chemical and biological character of the hyporheic zone may differ markedly from adjacent surface water and ground water.

Ground-water systems that discharge to streams can underlie extensive areas of the land surface (Figure 15). As a result, environmental conditions at the interface between ground water and surface water reflect changes in the broader landscape. For example, the types and numbers of organisms in a given reach of streambed result, in part, from interactions between water in the hyporheic zone and ground water from distant sources.

Figure 13. *The location where perennial streamflow begins in a channel can vary depending on the distribution of recharge in headwaters areas. Following dry periods (A), the start of streamflow will move up-channel during wet periods as the ground-water system becomes more saturated (B).*

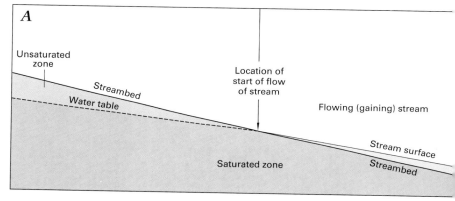

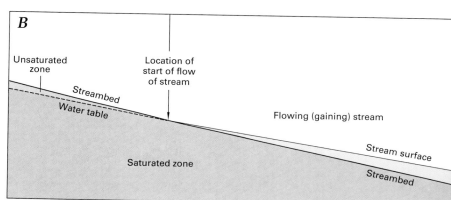

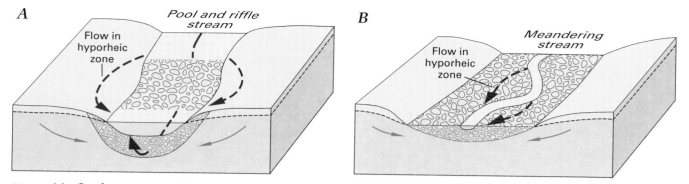

Figure 14. Surface-water exchange with ground water in the hyporheic zone is associated with abrupt changes in streambed slope (A) and with stream meanders (B).

Pool and riffle stream in Colorado. (Photograph by Robert Broshears.)

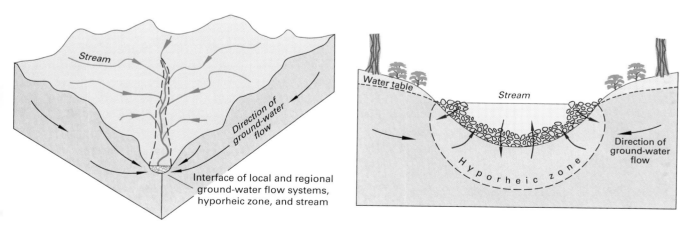

Figure 15. Streambeds and banks are unique environments because they are where ground water that drains much of the subsurface of landscapes interacts with surface water that drains much of the surface of landscapes.

INTERACTION OF GROUND WATER AND LAKES

Lakes interact with ground water in three basic ways: some receive ground-water inflow throughout their entire bed; some have seepage loss to ground water throughout their entire bed; but perhaps most lakes receive ground-water inflow through part of their bed and have seepage loss to ground water through other parts (Figure 16). Although these basic interactions are the same for lakes as they are for streams, the interactions differ in several ways.

The water level of natural lakes, that is, those not controlled by dams, generally does not change as rapidly as the water level of streams; therefore, bank storage is of lesser importance in lakes than it is in streams. Evaporation generally has a greater effect on lake levels than on stream levels because the surface area of lakes is generally larger and less shaded than many reaches of streams, and because lake water is not replenished as readily as a reach of a stream. Lakes can be present in many different parts of the landscape and can have complex ground-water flow systems associated with them. This is especially true for lakes in glacial and dune terrain, as is discussed in a later section of this Circular. Furthermore, lake sediments commonly have greater volumes of organic deposits than streams. These poorly permeable organic deposits can affect the distribution of seepage and biogeochemical exchanges of water and solutes more in lakes than in streams.

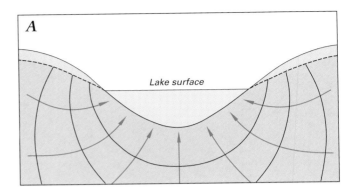

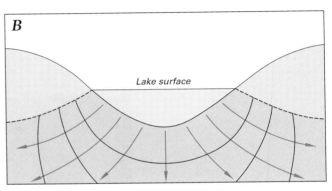

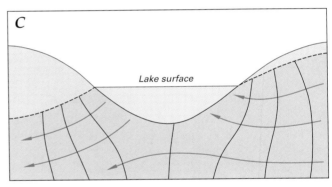

Figure 16. *Lakes can receive ground-water inflow (A), lose water as seepage to ground water (B), or both (C).*

Lake country in northern Wisconsin. (Photograph by David Krabbenhoft.)

Reservoirs are human-made lakes that are designed primarily to control the flow and distribution of surface water. Most reservoirs are constructed in stream valleys; therefore, they have some characteristics both of streams and lakes. Like streams, reservoirs can have widely fluctuating levels, bank storage can be significant, and they commonly have a continuous flushing of water through them. Like lakes, reservoirs can have significant loss of water by evaporation, significant cycling of chemical and biological materials within their waters, and extensive biogeochemical exchanges of solutes with organic sediments.

"Lakes and wetlands can receive ground-water inflow throughout their entire bed, have outflow throughout their entire bed, or have both inflow and outflow at different localities"

INTERACTION OF GROUND WATER AND WETLANDS

Wetlands are present in climates and landscapes that cause ground water to discharge to land surface or that prevent rapid drainage of water from the land surface. Similar to streams and lakes, wetlands can receive ground-water inflow, recharge ground water, or do both. Those wetlands that occupy depressions in the land surface have interactions with ground water similar to lakes and streams. Unlike streams and lakes, however, wetlands do not always occupy low points and depressions in the landscape (Figure 17A); they also can be present on slopes (such as fens) or even on drainage divides (such as some types of bogs). Fens are wetlands that commonly receive ground-water discharge (Figure 17B); therefore, they receive a continuous supply of chemical constituents dissolved in the ground water. Bogs are wetlands that occupy uplands (Figure 17D) or extensive flat areas, and they receive much of their water and chemical constituents from precipitation. The distribution of major wetland areas in the United States is shown in Figure 18.

In areas of steep land slopes, the water table sometimes intersects the land surface, resulting in ground-water discharge directly to the land surface. The constant source of water at these seepage faces (Figure 17B) permits the growth of wetland plants. A constant source of ground water to wetland plants is also provided to parts of the landscape that are downgradient from breaks in slope of the water table (Figure 17B), and where subsurface discontinuities in geologic units cause

Upland bog in Labrador, Canada. (Photograph by Lehn Franke.)

upward movement of ground water (Figure 17A). Many wetlands are present along streams, especially slow-moving streams. Although these riverine wetlands (Figure 17C) commonly receive ground-water discharge, they are dependent primarily on the stream for their water supply.

Wetlands in riverine and coastal areas have especially complex hydrological interactions because they are subject to periodic water-level changes. Some wetlands in coastal areas are affected by very predictable tidal cycles. Other coastal wetlands and riverine wetlands are more affected by seasonal water-level changes and by flooding. The combined effects of precipitation, evapotranspiration, and interaction with surface water and ground water result in a pattern of water depths in wetlands that is distinctive.

Hydroperiod is a term commonly used in wetland science that refers to the amplitude and frequency of water-level fluctuations. Hydroperiod affects all wetland characteristics, including the type of vegetation, nutrient cycling, and the types of invertebrates, fish, and bird species present.

Seepage face in Zion National Park, Utah. (Photograph by Robert Shedlock.)

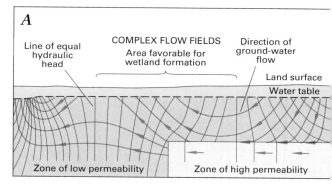

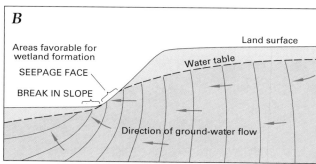

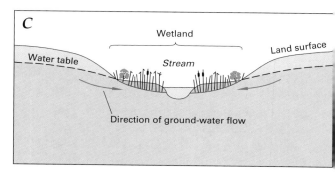

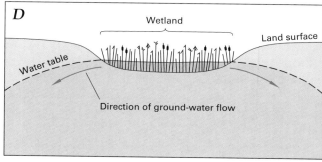

Figure 17. *The source of water to wetlands can be from ground-water discharge where the land surface is underlain by complex ground-water flow fields (A), from ground-water discharge at seepage faces and at breaks in slope of the water table (B), from streams (C), and from precipitation in cases where wetlands have no stream inflow and ground-water gradients slope away from the wetland (D).*

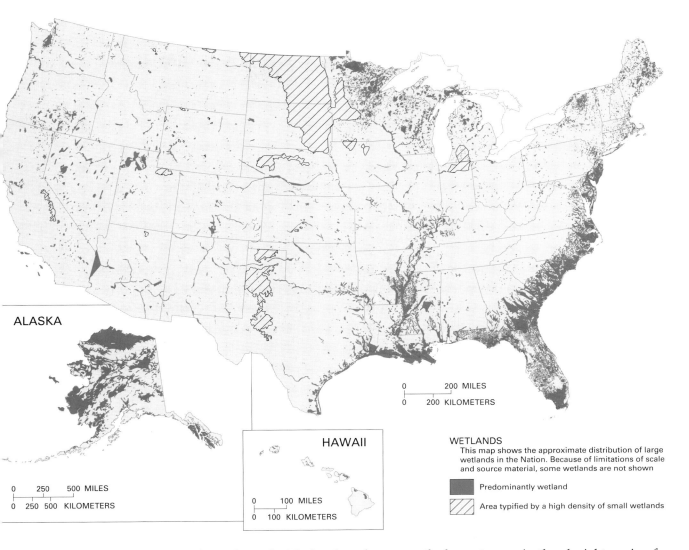

Figure 18. *Wetlands are present throughout the Nation, but they cover the largest areas in the glacial terrain of the north-central United States, coastal terrain along the Atlantic and gulf coasts, and riverine terrain in the lower Mississippi River Valley.*

A major difference between lakes and wetlands, with respect to their interaction with ground water, is the ease with which water moves through their beds. Lakes commonly are shallow around their perimeter where waves can remove fine-grained sediments, permitting the surface water and ground water to interact freely. In wetlands, on the other hand, if fine-grained and highly decomposed organic sediments are present near the wetland edge, the transfer of water and solutes between ground water and surface water is likely to be much slower.

Another difference in the interaction between ground water and surface water in wetlands compared to lakes is determined by rooted vegetation in wetlands. The fibrous root mat in wetland soils is highly conductive to water flow; therefore, water uptake by roots of emergent plants results in significant interchange between surface water and pore water of wetland sediments. The water exchanges in this upper soil zone even if exchange between surface water and ground water is restricted at the base of the wetland sediments.

Chemical Interactions of Ground Water and Surface Water

EVOLUTION OF WATER CHEMISTRY IN DRAINAGE BASINS

Two of the fundamental controls on water chemistry in drainage basins are the type of geologic materials that are present and the length of time that water is in contact with those materials. Chemical reactions that affect the biological and geochemical characteristics of a basin include (1) acid-base reactions, (2) precipitation and dissolution of minerals, (3) sorption and ion exchange, (4) oxidation-reduction reactions, (5) biodegradation, and (6) dissolution and exsolution of gases (see Box D). When water first infiltrates the land surface, microorganisms in the soil have a significant effect on the evolution of water chemistry. Organic matter in soils is degraded by microbes, producing high concentrations of dissolved carbon dioxide (CO_2). This process lowers the pH by increasing the carbonic acid (H_2CO_3) concentration in the soil water. The production of carbonic acid starts a number of mineral-weathering reactions, which result in bicarbonate (HCO_3^-) commonly being the most abundant anion in the water. Where contact times between water and minerals in shallow ground-water flow paths are short, the dissolved-solids concentration in the water generally is low. In such settings, limited chemical changes take place before ground water is discharged to surface water.

"Two of the fundamental controls on water chemistry in drainage basins are the type of geologic materials that are present and the length of time that water is in contact with those materials"

In deeper ground-water flow systems, the contact time between water and minerals is much longer than it is in shallow flow systems. As a result, the initial importance of reactions relating to microbes in the soil zone may be superseded over time by chemical reactions between minerals and water (geochemical weathering). As weathering progresses, the concentration of dissolved solids increases. Depending on the chemical composition of the minerals that are weathered, the relative abundance of the major inorganic chemicals dissolved in the water changes (see Box E).

Surface water in streams, lakes, and wetlands can repeatedly interchange with nearby ground water. Thus, the length of time water is in contact with mineral surfaces in its drainage basin can continue after the water first enters a stream, lake, or wetland. An important consequence of these continued interchanges between surface water and ground water is their potential to further increase the contact time between water and chemically reactive geologic materials.

CHEMICAL INTERACTIONS OF GROUND WATER AND SURFACE WATER IN STREAMS, LAKES, AND WETLANDS

Ground-water chemistry and surface-water chemistry cannot be dealt with separately where surface and subsurface flow systems interact. The movement of water between ground water and surface water provides a major pathway for chemical transfer between terrestrial and aquatic systems (see Box F). This transfer of chemicals affects the supply of carbon, oxygen, nutrients such as nitrogen and phosphorus, and other chemical constituents that enhance biogeochemical processes on both sides of the interface. This transfer can ultimately affect the biological and chemical characteristics of aquatic systems downstream.

"The movement of water between ground water and surface water provides a major pathway for chemical transfer between terrestrial and aquatic systems"

Some Common Types of Biogeochemical Reactions Affecting Transport of Chemicals in Ground Water and Surface Water

ACID-BASE REACTIONS

Acid-base reactions involve the transfer of hydrogen ions (H^+) among solutes dissolved in water, and they affect the effective concentrations of dissolved chemicals through changes in the H^+ concentration in water. A brief notation for H^+ concentration (activity) is pH, which represents a negative logarithmic scale of the H^+ concentration. Smaller values of pH represent larger concentrations of H^+, and larger values of pH represent smaller concentrations of H^+. Many metals stay dissolved when pH values are small; increased pH causes these metals to precipitate from solution.

PRECIPITATION AND DISSOLUTION OF MINERALS

Precipitation reactions result in minerals being formed (precipitated) from ions that are dissolved in water. An example of this type of reaction is the precipitation of iron, which is common in areas of ground-water seeps and springs. At these locations, the solid material iron hydroxide is formed when iron dissolved in ground water comes in contact with oxygen dissolved in surface water. The reverse, or dissolution reactions, result in ions being released into water by dissolving minerals. An example is the release of calcium ions (Ca^{++}) and bicarbonate ions (HCO_3^-) when calcite ($CaCO_3$) in limestone is dissolved.

SORPTION AND ION EXCHANGE

Sorption is a process in which ions or molecules dissolved in water (solutes) become attached to the surfaces (or near-surface parts) of solid materials, either temporarily or permanently. Thus, solutes in ground water and surface water can be sorbed either to the solid materials that comprise an aquifer or streambed or to particles suspended in ground water or surface water. The attachments of positively charged ions to clays and of pesticides to solid surfaces are examples of sorption. Release of sorbed chemicals to water is termed desorption.

When ions attached to the surface of a solid are replaced by ions that were in water, the process is known as ion exchange. Ion exchange is the process that takes place in water softeners; ions that contribute to water hardness—calcium and magnesium—are exchanged for sodium on the surface of the solid. The result of this process is that the amount of calcium and magnesium in the water declines and the amount of sodium increases. The opposite takes place when saltwater enters an aquifer; some of the sodium in the saltwater is exchanged for calcium sorbed to the solid material of the aquifer.

OXIDATION-REDUCTION REACTIONS

Oxidation-reduction (redox) reactions take place when electrons are exchanged among solutes. In these reactions, oxidation (loss of electrons) of certain elements is accompanied by the reduction (gain of electrons) of other elements.

For example, when iron dissolved in water that does not contain dissolved oxygen mixes with water that does contain dissolved oxygen, the iron and oxygen interact by oxidation and reduction reactions. The result of the reactions is that the dissolved iron loses electrons (the iron is oxidized) and oxygen gains electrons (the oxygen is reduced). In this case, the iron is an electron donor and the oxygen is an electron acceptor. Bacteria can use energy gained from oxidation-reduction reactions as they decompose organic material. To accomplish this, bacterially mediated oxidation-reduction reactions use a sequence of electron acceptors, including oxygen, nitrate, iron, sulfate, and carbon dioxide. The presence of the products of these reactions in ground water and surface water can be used to identify the dominant oxidation-reduction reactions that have taken place in those waters. For example, the bacterial reduction of sulfate (SO_4^{2-}) to sulfide (HS^-) can result when organic matter is oxidized to CO_2.

BIODEGRADATION

Biodegradation is the decomposition of organic chemicals by living organisms using enzymes. Enzymes are specialized organic compounds made by living organisms that speed up reactions with other organic compounds. Microorganisms degrade (transform) organic chemicals as a source of energy and carbon for growth. Microbial processes are important in the fate and transport of many organic compounds. Some compounds, such as petroleum hydrocarbons, can be used directly by microorganisms as food sources and are rapidly degraded in many situations. Other compounds, such as chlorinated solvents, are not as easily assimilated. The rate of biodegradation of an organic chemical is dependent on its chemical structure, the environmental conditions, and the types of microorganisms that are present. Although biodegradation commonly can result in complete degradation of organic chemicals to carbon dioxide, water, and other simple products, it also can lead to intermediate products that are of environmental concern. For example, deethylatrazine, an intermediate degradation product of the pesticide atrazine (see Box P), commonly is detected in water throughout the corn-growing areas of the United States.

DISSOLUTION AND EXSOLUTION OF GASES

Gases are directly involved in many geochemical reactions. One of the more common gases is carbon dioxide (CO_2). For example, stalactites can form in caves when dissolved CO_2 exsolves (degasses) from dripping ground water, causing pH to rise and calcium carbonate to precipitate. In soils, the microbial production of CO_2 increases the concentration of carbonic acid (H_2CO_3), which has a major control on the solubility of aquifer materials. Other gases commonly involved in chemical reactions are oxygen, nitrogen, hydrogen sulfide (H_2S), and methane (CH_4). Gases such as chlorofluorocarbons (CFCs) and radon are useful as tracers to determine the sources and rates of ground-water movement (see Box G).

Evolution of Ground-Water Chemistry from Recharge to Discharge Areas in the Atlantic Coastal Plain

Changes in the chemical composition of ground water in sediments of the Atlantic Coastal Plain (Figure E–1) provide an example of the chemical evolution of ground water in a regional flow system. In the shallow regime, infiltrating water comes in contact with gases in the unsaturated zone and shallow ground water. As a result of this contact, localized, short-term, fast reactions take place that dissolve minerals and degrade organic material. In the deep regime, long-term, slower chemical reactions, such as precipitation and dissolution of minerals and ion-exchange, add or remove solutes. These natural processes and reactions commonly produce a predictable sequence of hydrochemical facies. In the Atlantic Coastal Plain, ground water evolves from water containing abundant bicarbonate ions and small concentrations of dissolved solids near the point of recharge to water containing abundant chloride ions and large concentrations of dissolved solids where it discharges into streams, estuaries, and the Atlantic Ocean.

Figure E–1. In a coastal plain, such as along the Atlantic Coast of the United States, the interrelations of different rock types, shallow and deep ground-water flow systems (regimes), and mixing with saline water (A) results in the evolution of a number of different ground-water chemical types (B). (Modified from Back, William, Baedecker, M.J., and Wood, W.W., 1993, Scales in chemical hydrogeology—A historical perspective, in Alley, W.M., ed., Regional Ground-Water Quality: New York, van Nostrand Reinhold, p. 111–129.) (Reprinted by permission of John Wiley & Sons, Inc.)

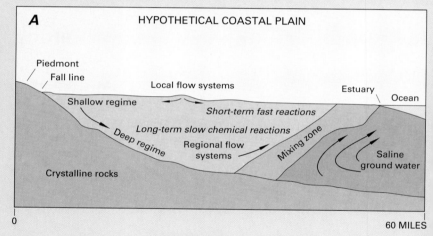

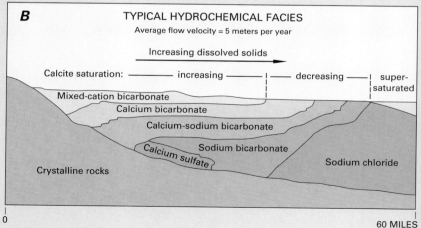

Many streams are contaminated. Therefore, the need to determine the extent of the chemical reactions that take place in the hyporheic zone is widespread because of the concern that the contaminated stream water will contaminate shallow ground water (see Box G). Streams offer good examples of how interconnections between ground water and surface water affect chemical processes. Rough channel bottoms cause stream water to enter the streambed and to mix with ground water in the hyporheic zone. This mixing establishes sharp changes in chemical concentrations in the hyporheic zone.

A zone of enhanced biogeochemical activity usually develops in shallow ground water as a result of the flow of oxygen-rich surface water into the subsurface environment, where bacteria and geochemically active sediment coatings are abundant (Figure 19). This input of oxygen to the streambed stimulates a high level of activity by aerobic (oxygen-using) microorganisms if dissolved oxygen is readily available. It is not uncommon for dissolved oxygen to be completely used up in hyporheic flow paths at some distance into the streambed, where anaerobic microorganisms dominate microbial activity. Anaerobic bacteria can use nitrate, sulfate, or other solutes in place of oxygen for metabolism. The result of these processes is that many solutes are highly reactive in shallow ground water in the vicinity of streambeds.

The movement of nutrients and other chemical constituents, including contaminants, between ground water and surface water is affected by biogeochemical processes in the hyporheic zone. For example, the rate at which organic contaminants biodegrade in the hyporheic zone can exceed rates in stream water or in ground water away from the stream. Another example is the removal of dissolved metals in the hyporheic zone. As water passes through the hyporheic zone, dissolved metals are removed by precipitation of metal oxide coatings on the sediments.

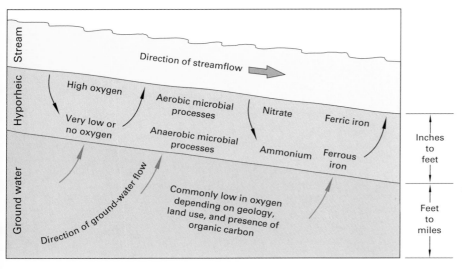

Figure 19. *Microbial activity and chemical transformations commonly are enhanced in the hyporheic zone compared to those that take place in ground water and surface water. This diagram illustrates some of the processes and chemical transformations that may take place in the hyporheic zone. Actual chemical interactions depend on numerous factors including aquifer mineralogy, shape of the aquifer, types of organic matter in surface water and ground water, and nearby land use.*

The Interface Between Ground Water and Surface Water as an Environmental Entity

In the bed and banks of streams, water and solutes can exchange in both directions across the streambed. This process, termed hyporheic exchange, creates subsurface environments that have variable proportions of water from ground water and surface water. Depending on the type of sediment in the streambed and banks, the variability in slope of the streambed, and the hydraulic gradients in the adjacent ground-water system, the hyporheic zone can be as much as several feet in depth and hundreds of feet in width. The dimensions of the hyporheic zone generally increase with increasing width of the stream and permeability of streambed sediments.

The importance of the hyporheic zone was first recognized when higher than expected abundances of aquatic insects were found in sediments where concentrations of oxygen were high. Caused by stream-water input, the high oxygen concentrations in the hyporheic zone make it possible for organisms to live in the pore spaces in the sediments, thereby providing a refuge for those organisms. Also, spawning success of salmon is greater where flow from the stream brings oxygen into contact with eggs that were deposited within the coarse sediment.

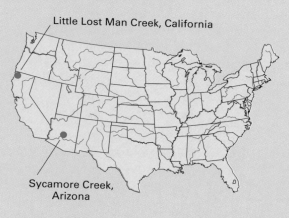

The hyporheic zone also can be a source of nutrients and algal cells to streams that foster the recovery of streams following catastrophic storms. For example, in a study of the ecology of Sycamore Creek in Arizona, it was found that the algae that grew in the top few inches of streambed sediment were quickest to recover following storms in areas where water in the sediments moved upward (Figure F–1).

These algae recovered rapidly following storms because concentrations of dissolved nitrogen were higher in areas of the streambed where water moved upward than in areas where water moved downward. Areas of streambed where water moved upward are, therefore, likely to be the first areas to return to more normal ecological conditions following flash floods in desert streams.

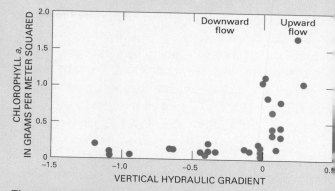

Figure F–1. Abundance of algae in streambed sediments, as indicated by concentration of chlorophyll a, was markedly greater in areas where water moved upward through the sediments than in areas where water moved downward through the sediments in Sycamore Creek in Arizona. (Modified from Valett, H.M., Fisher, S.G., Grimm, N.B., and Camill, P., 1994, Vertical hydrologic exchange and ecologic stability of a desert stream ecosystem: Ecology, v. 75, p. 548–560.) (Reprinted with permission.)

Sycamore Creek, Arizona. (Photograph by Owen Baynham.)

Hyporheic zones also serve as sites for nutrient uptake. A study of a coastal mountain stream in northern California indicated that transport of dissolved oxygen, dissolved carbon, and dissolved nitrogen in stream water into the hyporheic zone stimulated uptake of nitrogen by microbes and algae attached to sediment. A model simulation of nitrogen uptake (Figure F–2) indicated that both the physical process of water exchange between the stream and the hyporheic zone and the biological uptake of nitrate in the hyporheic zone affected the concentration of dissolved nitrogen in the stream.

The importance of biogeochemical processes that take place at the interface of ground water and surface water in improving water quality for human consumption is shown by the following example. Decreasing metal concentrations (Figure F–3) in drinking-water wells adjacent to the River Glatt in Switzerland was attributed to the interaction of the river with subsurface water. The improvement in ground-water quality started with improved sewage-treatment plants, which lowered phosphate in the river. Lower phosphate concentrations lowered the amount of algal production in the river, which decreased the amount of dissolved organic carbon flowing into the riverbanks. These factors led to a decrease in the bacteria-caused dissolution of manganese and cadmium that were present as coatings on sediment in the aquifer. The result was substantially lower dissolved metal concentrations in ground water adjacent to the river, which resulted in an unexpected improvement in the quality of drinking water.

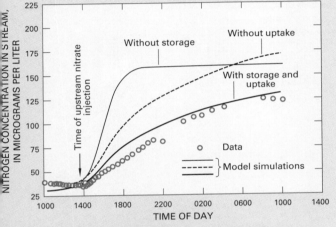

Figure F–2. Nitrate injected into Little Lost Man Creek in northern California was stored and taken up by algae and microbes in the hyporheic zone. (Modified from Kim, B.K.A., Jackman, A.P., and Triska, F.J., 1992, Modeling biotic uptake by periphyton and transient hyporheic storage of nitrate in a natural stream: Water Resources Research, v. 28, no.10, p. 2743–2752.)

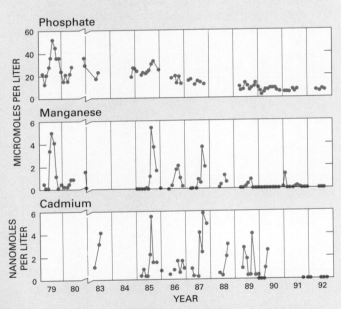

Figure F–3. A decline in manganese and cadmium concentrations after 1990 in drinking-water wells near the River Glatt in Switzerland was attributed to decreased phosphate in the river and hydrologic and biogeochemical interactions between river water and ground water. (Modified from von Gunten, H.R., and Lienert, Ch., 1993, Decreased metal concentrations in ground water caused by controls on phosphate emissions: Nature, v. 364, p. 220–222.) (Reprinted with permission from Nature, Macmillan Magazines Limited.)

Little Lost Man Creek, California. (Photograph by Judson Harvey.)

Use of Environmental Tracers to Determine the Interaction of Ground Water and Surface Water

Environmental tracers are naturally occurring dissolved constituents, isotopes, or physical properties of water that are used to track the movement of water through watersheds. Useful environmental tracers include (1) common dissolved constituents, such as major cations and anions; (2) stable isotopes of oxygen (^{18}O) and hydrogen (^{2}H) in water molecules; (3) radioactive isotopes such as tritium (^{3}H) and radon (^{222}Rn); and (4) water temperature. When used in simple hydrologic transport calculations, environmental tracers can be used to (1) determine source areas of water and dissolved chemicals in drainage basins, (2) calculate hydrologic and chemical fluxes between ground water and surface water, (3) calculate water ages that indicate the length of time water and dissolved chemicals have been present in the drainage basin (residence times), and (4) determine average rates of chemical reactions that take place during transport. Some examples are described below.

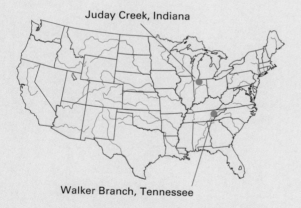

Major cations and anions have been used as tracers in studies of the hydrology of small watersheds to determine the sources of water to streamflow during storms (see Figure G–1). In addition, stable isotopes of oxygen and hydrogen, which are part of water molecules, are useful for determining the mixing of waters from different source areas because of such factors as (1) differences in the isotopic composition of precipitation among recharge areas, (2) changes in the isotopic composition of shallow subsurface water caused by evaporation, and (3) temporal variability in the isotopic composition of precipitation relative to ground water.

Radioactive isotopes are useful indicators of the time that water has spent in the ground-water system. For example, tritium (^{3}H) is a well-known radioactive isotope of hydrogen that had peak concentrations in precipitation in the mid-1960s as a result of above-ground nuclear-bomb testing conducted at that time. Chlorofluorocarbons (CFCs), which are industrial chemicals that are present in ground water less than 50 years old, also can be used to calculate ground-water age in different parts of a drainage basin.

^{222}Radon is a chemically inert, radioactive gas that has a half-life of only 3.83 days. It is produced naturally in ground water as a product of the radioactive decay of ^{226}radium in uranium-bearing rocks and sediment. Several studies have documented that radon can be used to identify locations of

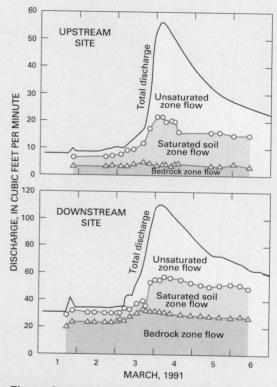

Figure G–1. The relative contributions of different subsurface water sources to streamflow in a stream in Tennessee were determined by analyzing the relative concentrations of calcium and sulfate. Note that increases in bedrock zone (ground water) flow appear to contribute more to the stormflow response at the downstream site than to the stormflow response at the upstream site in this small watershed. (Modified from Mulholland, P.J., 1993, Hydrometric and stream chemistry evidence of three storm flowpaths in Walker Branch Watershed: Journal of Hydrology, v. 151, p. 291–316.) (Reprinted with permission from Elsevier Science-NL, Amsterdam, The Netherlands.)

significant ground-water input to a stream, such as from springs. Radon also has been used to determine stream-water movement to ground water. For example, radon was used in a study in France to determine stream-water loss to ground water as a result of ground-water withdrawals. (See Figure G–2.)

An example of using stream-water temperature and sediment temperature for mapping gaining and losing reaches of a stream is shown in Figure G–3. In gaining reaches of the stream, sediment temperature and stream-water temperature are markedly different. In losing reaches of the stream, the diurnal fluctuations of temperature in the stream are reflected more strongly in the sediment temperature.

Sampling for a chemical tracer. (Photograph by Gary Zellweger.)

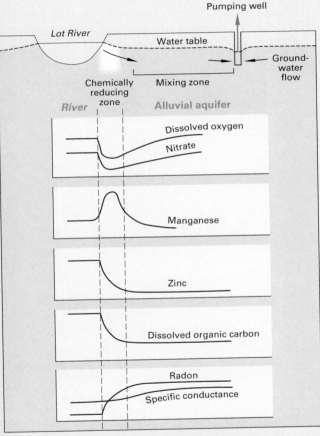

Figure G–2. Sharp changes in chemical concentrations were detected over short distances as water from the Lot River in France moved into its contiguous alluvial aquifer in response to pumping from a well. Specific conductance of water was used as an environmental tracer to determine the extent of mixing of surface water with ground water, and radon was used to determine the inflow rate of stream water. Both pieces of information were then used to calculate the rate at which dissolved metals reacted to form solid phases during movement of stream water toward the pumping well. (Modified from Bourg, A.C.M., and Bertin, C., 1993, Biogeochemical processes during the infiltration of river water into an alluvial aquifer: Environmental Science and Technology, v. 27, p. 661–666.) (Reprinted with permission from the American Chemical Society.)

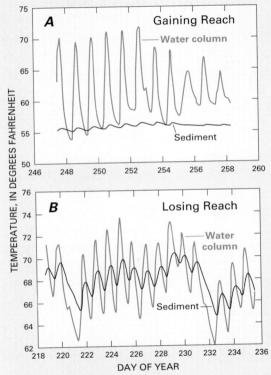

Figure G–3. Ground-water temperatures generally are more stable than surface-water temperatures. Therefore, gaining reaches of Juday Creek in Indiana are characterized by relatively stable sediment temperatures compared to stream-water temperatures (A). Conversely, losing reaches are characterized by more variable sediment temperatures caused by the temperature of the inflowing surface water (B). (Modified from Silliman, S.E., and Booth, D.F., 1993, Analysis of time series measurements of sediment temperature for identification of gaining versus losing portions of Juday Creek, Indiana: Journal of Hydrology, v. 146, p. 131–148.) (Reprinted with permission from Elsevier Science-NL, Amsterdam, The Netherlands.)

Lakes and wetlands also have distinctive biogeochemical characteristics with respect to their interaction with ground water. The chemistry of ground water and the direction and magnitude of exchange with surface water significantly affect the input of dissolved chemicals to lakes and wetlands. In general, if lakes and wetlands have little interaction with streams or with ground water, input of dissolved chemicals is mostly from precipitation; therefore, the input of chemicals is minimal. Lakes and wetlands that have a considerable amount of ground-water inflow generally have large inputs of dissolved chemicals. In cases where the input of dissolved nutrients such as phosphorus and nitrogen exceeds the output, primary production by algae and wetland plants is large. When this large amount of plant material dies, oxygen is used in the process of decomposition. In some cases the loss of oxygen from lake water can be large enough to kill fish and other aquatic organisms.

The magnitude of surface-water inflow and outflow also affects the retention of nutrients in wetlands. If lakes or wetlands have no stream outflow, retention of chemicals is high. The tendency to retain nutrients usually is less in wetlands that are flushed substantially by throughflow of surface water. In general, as surface-water inputs increase, wetlands vary from those that strongly retain nutrients to those that both import and export large amounts of nutrients. Furthermore, wetlands commonly have a significant role in altering the chemical form of dissolved constituents. For example, wetlands that have throughflow of surface water tend to retain the chemically oxidized forms and release the chemically reduced forms of metals and nutrients.

Eutrophic lake in Saskatchewan, Canada. (Photograph by James LaBaugh.)

"The chemistry of ground water and the direction and magnitude of exchange with surface water significantly affect the input of dissolved chemicals to lakes and wetlands"

Interaction of Ground Water and Surface Water in Different Landscapes

Ground water is present in virtually all landscapes. The interaction of ground water with surface water depends on the physiographic and climatic setting of the landscape. For example, a stream in a wet climate might receive ground-water inflow, but a stream in an identical physiographic setting in an arid climate might lose water to ground water. To provide a broad and unified perspective of the interaction of ground water and surface water in different landscapes, a conceptual landscape (Figure 2) is used as a reference. Some common features of the interaction for various parts of the conceptual landscape are described below. The five general types of terrain discussed are mountainous, riverine, coastal, glacial and dune, and karst.

MOUNTAINOUS TERRAIN

The hydrology of mountainous terrain (area M of the conceptual landscape, Figure 2) is characterized by highly variable precipitation and water movement over and through steep land slopes. On mountain slopes, macropores created by burrowing organisms and by decay of plant roots have the capacity to transmit subsurface flow downslope quickly. In addition, some rock types underlying soils may be highly weathered or fractured and may transmit significant additional amounts of flow through the subsurface. In some settings this rapid flow of water results in hillside springs.

A general concept of water flow in mountainous terrain includes several pathways by which precipitation moves through the hillside to a stream (Figure 20). Between storm and snowmelt periods, streamflow is sustained by discharge from the ground-water system (Figure 20A). During intense storms, most water reaches streams very rapidly by partially saturating and flowing through the highly conductive soils. On the lower parts of hillslopes, the water table sometimes rises to the land surface during storms, resulting in overland flow (Figure 20B). When this occurs, precipitation on the saturated area adds to the quantity of overland flow. When storms or snowmelt persist in mountainous areas, near-stream saturated areas can expand outward from streams to include areas higher on the hillslope. In some settings, especially in arid regions, overland flow can be generated when the rate of rainfall exceeds the infiltration capacity of the soil (Figure 20C).

Near the base of some mountainsides, the water table intersects the steep valley wall some distance up from the base of the slope (Figure 21, left side of valley). This results in perennial discharge of ground water and, in many cases,

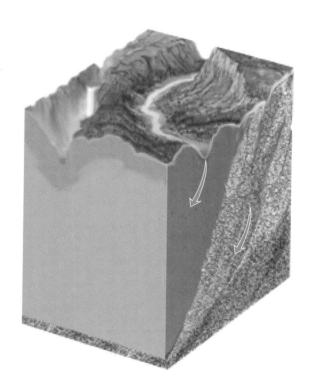

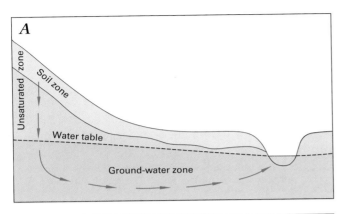

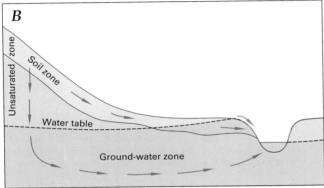

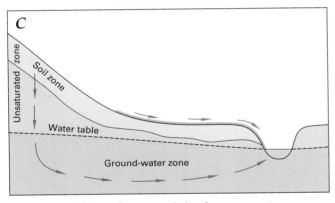

Figure 20. *Water from precipitation moves to mountain streams along several pathways. Between storms and snowmelt periods, most inflow to streams commonly is from ground water (A). During storms and snowmelt periods, much of the water inflow to streams is from shallow flow in saturated macropores in the soil zone. If infiltration to the water table is large enough, the water table will rise to the land surface and flow to the stream is from ground water, soil water, and overland runoff (B). In arid areas where soils are very dry and plants are sparse, infiltration is impeded and runoff from precipitation can occur as overland flow (C). (Modified from Dunne, T., and Leopold, L.B., 1978, Water in environmental planning: San Francisco, W.H. Freeman.) (Used with permission.)*

the presence of wetlands. A more common hydrologic process that results in the presence of wetlands in some mountain valleys is the upward discharge of ground water caused by the change in slope of the water table from being steep on the valley side to being relatively flat in the alluvial valley (Figure 21, right side of valley). Where both of these water-table conditions exist, wetlands fed by ground water, which commonly are referred to as fens, can be present.

Another dynamic aspect of the interaction of ground water and surface water in mountain settings is caused by the marked longitudinal component of flow in mountain valleys. The high gradient of mountain streams, coupled with the coarse texture of streambed sediments, results in a strong down-valley component of flow accompanied by frequent exchange of stream water with water in the hyporheic zone (Figure 14) (see Box H). The driving force for water exchange between a stream and its hyporheic zone is created by the surface water flowing over rough streambeds, through pools and riffles, over cascades, and around boulders and logs. Typically, the stream enters the hyporheic zone at the downstream end of pools and then flows beneath steep sections of the stream (called riffles), returning to the stream at the upstream end of the next pool (Figure 14A). Stream water also may enter the hyporheic zone upstream from channel meanders, causing stream water to flow through a gravel bar before reentering the channel downstream (Figure 14B).

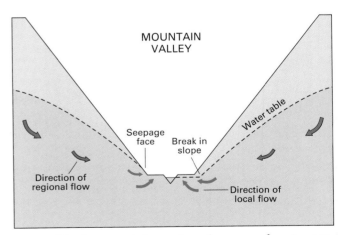

Figure 21. *In mountainous terrain, ground water can discharge at the base of steep slopes (left side of valley), at the edges of flood plains (right side of valley), and to the stream.*

Streams flowing from mountainous terrain commonly flow across alluvial fans at the edges of the valleys. Most streams in this type of setting lose water to ground water as they traverse the highly permeable alluvial fans. This process has long been recognized in arid western regions, but it also has been documented in humid regions, such as the Appalachian Mountains. In arid and semiarid regions, seepage of water from the stream can be the principal source of aquifer recharge. Despite its importance, ground-water

Alluvial fan in Alaska. (Photograph by Earl Brabb.)

Mountain stream in Oregon. (Photograph by Dennis Wentz.)

recharge from losing streams remains a highly uncertain part of the water balance of aquifers in these regions. Promising new methods of estimating ground-water recharge, at least locally, along mountain fronts are being developed—these methods include use of environmental tracers, measuring vertical temperature profiles in streambeds, measuring hydraulic characteristics of streambeds, and measuring the difference in hydraulic head between the stream and the underlying aquifer.

The most common natural lakes in mountainous terrain are those that are dammed by rock sills or glacial deposits high in the mountains.

Termed cirque lakes, they receive much of their water from snowmelt. However, they interact with ground water much like the processes shown in Figure 21, and they can be maintained by ground water throughout the snow-free season.

The geochemical environment of mountains is quite diverse because of the effects of highly variable climate and many different rock and soil types on the evolution of water chemistry. Geologic materials can include crystalline, volcanic, and sedimentary rocks and glacial deposits. Sediments can vary from those having well-developed soil horizons to stream alluvium that has no soil development. During heavy precipitation, much water flows through shallow flow paths, where it interacts with microbes and soil gases. In the deeper flow through fractured bedrock, longer term geochemical interactions of ground water with minerals determine the chemistry of water that eventually discharges to streams. Base flow of streams in mountainous terrain is derived by drainage from saturated alluvium in valley bottoms and from drainage of bedrock fractures. Mixing of these chemically different water types results in geochemical reactions that affect the chemistry of water in streams. During downstream transport in the channel, stream water mixes with ground water in the hyporheic zone. In some mountain streams, the volume of water in the hyporheic zone is considerably larger than that in the stream channel. Chemical reactions in hyporheic zones can, in some cases, substantially alter the water chemistry of streams (Figure 19).

Field Studies of Mountainous Terrain

The steep slopes and rocky characteristics of mountainous terrain make it difficult to determine interactions of ground water and surface water. Consequently, few detailed hydrogeologic investigations of these interactions have been conducted in mountainous areas. Two examples are given below.

A field and modeling study of the Mirror Lake area in the White Mountains of New Hampshire indicated that the sizes of ground-water flow systems contributing to surface-water bodies were considerably larger than their topographically defined watersheds. For example, much of the ground water in the fractured bedrock that discharges to Mirror Lake passes beneath the local flow system associated with Norris Brook (Figure H–1). Furthermore, a more extensive deep ground-water flow system that discharges to the Pemigewasset River passes beneath flow systems associated with both Norris Brook and Mirror Lake.

Studies in mountainous terrain have used tracers to determine sources of ground water to streams (see Box G). In addition to revealing processes of water exchange between ground water and stream water, solute tracers have proven useful for defining the limits of the hyporheic zone surrounding mountain streams. For example, solute tracers such as chloride or bromide ions are injected into the stream to artificially raise concentrations above natural background concentrations. The locations and amounts of ground-water inflow are determined from a simple dilution model. The extent that tracers move into the hyporheic zone can be estimated by the models and commonly is verified by sampling wells placed in the study area.

Mirror Lake, New Hampshire. (Photograph by Thomas Winter.)

Figure H–1. Ground-water flow systems in the Mirror Lake area extend beyond the topographically defined surface-water watersheds. (Modified from Harte, P.T., and Winter, T.C., 1996, Factors affecting recharge to crystalline rock in the Mirror Lake area, Grafton County, New Hampshire: in Morganwalp, D.W., and Aronson, D.A., eds., U.S. Geological Survey Toxic Substances Hydrology Program—Proceedings of Technical Meeting, Colorado Springs, Colorado, September 20–24, 1993: U.S. Geological Survey Water-Resources Investigations Report 94–4014, p. 141–150.)

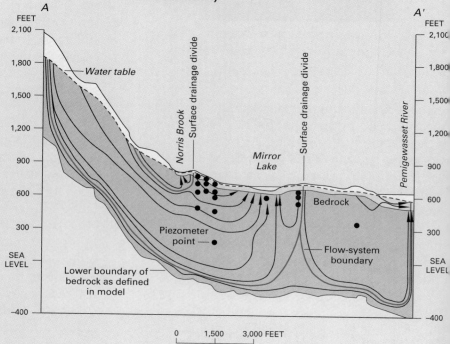

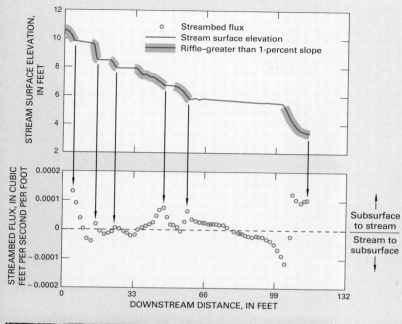

Figure H–2. In mountain streams characterized by pools and riffles, such as at Saint Kevin Gulch in Colorado, inflow of water from the hyporheic zone to the stream was greatest at the downstream end of riffles. (Modified from Harvey, J.W., and Bencala, K.E., 1993, The effect of streambed topography on surface-subsurface water exchange in mountain catchments: Water Resources Research, v. 29, p. 89–98.)

Saint Kevin Gulch, Colorado. (Photograph by Kenneth Bencala.)

Chalk Creek, Colorado. (Photograph by Briant Kimball.)

A study in Colorado indicated that hyporheic exchange in mountain streams is caused to a large extent by the irregular topography of the streambed, which creates pools and riffles characteristic of mountain streams. Ground water enters streams most readily at the upstream end of deep pools, and stream water flows into the subsurface beneath and to the side of steep sections of streams (riffles) (Figure H–2). Channel irregularity, therefore, is an important control on the location of ground-water inflow to streams and on the size of the hyporheic zone in mountain streams because changes in slope determine the length and depth of hyporheic flow paths.

The source and fate of metal contaminants in streams receiving drainage from abandoned mines can be determined by using solute tracers. In addition to surface drainage from mines, a recent study of Chalk Creek in Colorado indicated that contaminants were being brought to the stream by ground-water inflow. The ground water had been contaminated from mining activities in the past and is now a new source of contamination to the stream. This nonpoint ground-water source of contamination will very likely be much more difficult to clean up than the point source of contamination from the mine tunnel.

RIVERINE TERRAIN

In some landscapes, stream valleys are small and they commonly do not have well-developed flood plains (area R of the conceptual landscape, Figure 2) (see Box I). However, major rivers (area V of the reference landscape, Figure 2) have valleys that usually become increasingly wider downstream. Terraces, natural levees, and abandoned river meanders are common landscape features in major river valleys, and wetlands and lakes commonly are associated with these features.

The interaction of ground water and surface water in river valleys is affected by the interchange of local and regional ground-water flow systems with the rivers and by flooding and evapotranspiration. Small streams receive ground-water inflow primarily from local flow systems, which usually have limited extent and are highly variable seasonally. Therefore, it is not unusual for small streams to have gaining or losing reaches that change seasonally.

For larger rivers that flow in alluvial valleys, the interaction of ground water and surface water usually is more spatially diverse than it is for smaller streams. Ground water from regional flow systems discharges to the river as well as at various places across the flood plain (Figure 22). If terraces are present in the alluvial valley, local ground-water flow systems may be associated with each terrace, and lakes and wetlands may be formed because of this source of ground water. At some locations, such as at the valley wall and at the river, local and regional ground-water flow systems may discharge in close proximity. Furthermore, in large alluvial valleys, significant down-valley components of flow in the streambed and in the shallow alluvium also may be present (see Box I).

Alluvial valley of the Mississippi River. (Photograph by Robert Meade.)

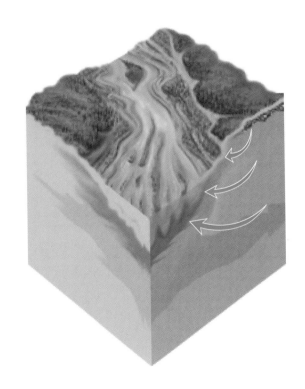

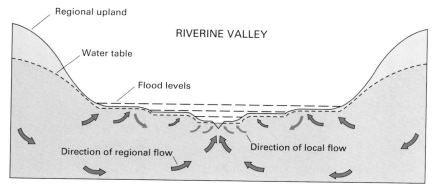

Figure 22. *In broad river valleys, small local ground-water flow systems associated with terraces overlie more regional ground-water flow systems. Recharge from flood waters superimposed on these ground-water flow systems further complicates the hydrology of river valleys.*

Added to this distribution of ground-water discharge from different flow systems to different parts of the valley is the effect of flooding. At times of high river flows, water moves into the ground-water system as bank storage (Figure 11). The flow paths can be as lateral flow through the riverbank (Figure 12B) or, during flooding, as vertical seepage over the flood plain (Figure 12C). As flood waters rise, they cause bank storage to move into higher and higher terraces.

The water table generally is not far below the land surface in alluvial valleys. Therefore, vegetation on flood plains, as well as at the base of some terraces, commonly has root systems deep enough so that the plants can transpire water directly from ground water. Because of the relatively stable source of ground water, particularly in areas of ground-water discharge, the vegetation can transpire water near the maximum potential transpiration rate, resulting in the same effect as if the water were being pumped by a well (see Figure 7). This large loss of water can result in drawdown of the water table such that the plants intercept some of the water that would otherwise flow to the river, wetland, or lake. Furthermore, in some settings it is not uncommon during the growing season for the pumping effect of transpiration to be significant enough that surface water moves into the subsurface to replenish the transpired ground water.

Riverine alluvial deposits range in size from clay to boulders, but in many alluvial valleys, sand and gravel are the predominant deposits. Chemical reactions involving dissolution or precipitation of minerals (see Box D) commonly do not have a significant effect on water chemistry in sand and gravel alluvial aquifers because the rate of water movement is relatively fast compared to weathering rates. Instead, sorption and desorption reactions and oxidation/reduction reactions related to the activity of microorganisms probably have a greater effect on water chemistry in these systems. As in small streams, biogeochemical processes in the hyporheic zone may have a significant effect on the chemistry of ground water and surface water in larger riverine systems. Movement of oxygen-rich surface water into the subsurface, where chemically reactive sediment coatings are abundant, causes increased chemical reactions related to activity of microorganisms. Sharp gradients in concentration of some chemical constituents in water, which delimit this zone of increased biogeochemical activity, are common near the boundary between ground water and surface water. In addition, chemical reactions in the hyporheic zone can cause precipitation of some reactive solutes and contaminants, thereby affecting water quality.

Field Studies of Riverine Terrain

Streams are present in virtually all landscapes, and in some landscapes, they are the principal surface-water features. The interaction of ground water with streams varies in complexity because they vary in size from small streams near headwaters areas to large rivers flowing in large alluvial valleys, and also because streams intersect ground-water flow systems of greatly different scales. Examples of the interaction of ground water and surface water for small and large riverine systems are presented below.

The Straight River, which runs through a sand plain in central Minnesota, is typical of a small stream that does not have a flood plain and that derives most of its water from ground-water inflow. The water-table contours near the river bend sharply upstream (Figure I–1), indicating that ground water moves directly into the river. It is estimated from base-flow studies (see Box B) that, on an annual basis, ground water accounts for more than 90 percent of the water in the river.

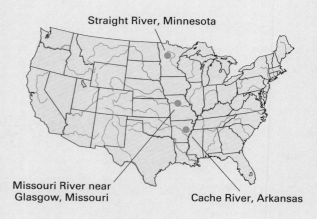

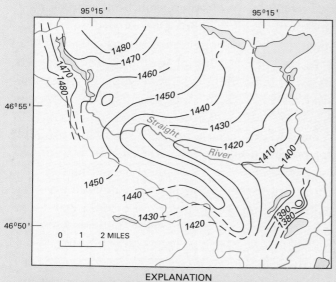

EXPLANATION
—1420— WATER-TABLE CONTOUR—Shows altitude of the water table in feet above sea level. Dashed where approximately located. Contour interval 10 feet

Figure I–1. *Small streams, such as the Straight River in Minnesota, commonly do not have flood plains. The flow of ground water directly into the river is indicated by the water-table contours that bend sharply upstream. (Modified from Stark, J.R., Armstrong, D.S., and Zwilling, D.R., 1994, Stream-aquifer interactions in the Straight River area, Becker and Hubbard Counties, Minnesota: U.S. Geological Survey Water-Resources Investigations Report 94–4009, 83 p.)*

Straight River, Minnesota. (Photograph by James Stark.)

In contrast, the results of a study of the lower Missouri River Valley indicate the complexity of ground-water flow and its interaction with streams in large alluvial valleys. Configuration of the water table in this area indicates that ground water flows into the river at right angles in some reaches, and it flows parallel to the river in others (Figure I–2A). This study also resulted in a map that showed patterns of water-table fluctuations with respect to proximity to the river (Figure I–2B). This example shows the wide variety of ground-water flow conditions that can be present in large alluvial valleys.

Another study of part of a large alluvial valley provides an example of the presence of smaller scale flow conditions. The Cache River is a stream within the alluvial valley of the Mississippi River Delta system in eastern Arkansas. In a study of the Black Swamp, which lies along a reach of the river, a number of wells and piezometers were installed to determine the interaction of ground water with the swamp and the river. By measuring hydraulic head at different depths in the

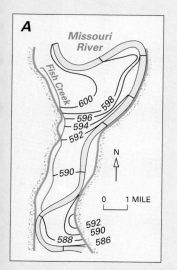
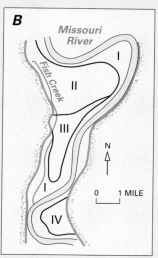

EXPLANATION

—590— WATER-TABLE CONTOUR—Shows altitude of water table in feet above sea level. Contour interval 2 feet

Figure I–2. In flood plains of large rivers, such as the Missouri River near Glasgow, Missouri, patterns of ground-water movement (A) and water-table fluctuations (B) can be complex. Zone I is an area of rapidly fluctuating water levels, zone II is an area of long-term stability, zone III is an area of down-valley flow, and zone IV is a persistent ground-water high. (Modified from Grannemann, N.G., and Sharp, J.M., Jr., 1979, Alluvial hydrogeology of the lower Missouri River: Journal of Hydrology, v. 40, p. 85–99.) (Reprinted with permission from Elsevier Science-NL, Amsterdam, The Netherlands.)

alluvium, it was possible to construct a hydrologic section through the alluvium (Figure I–3), showing that the river receives ground-water discharge from both local and regional ground-water flow systems. In addition, the section also shows the effect of the break in slope associated with the terrace at the edge of the swamp, which causes ground water from a local flow system to discharge into the edge of the swamp rather than to the river.

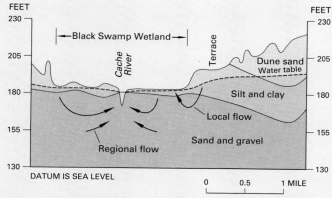

Figure I–3. The Cache River in Arkansas provides an example of contributions to a river from regional and local ground-water flow systems. In addition, a small local ground-water flow system associated with a terrace discharges to the wetland at the edge of the flood plain. (Modified from Gonthier, G.J., 1996, Ground-water flow conditions within a bottomland hardwood wetland, eastern Arkansas: Wetlands, v. 16, no. 3, p. 334–346.) (Used with permission.)

Missouri River Valley, near Atchison, Kansas. (Photograph by Robert Meade.)

Cache River, Arkansas. (Photograph by Gerard Gonthier.)

COASTAL TERRAIN

Coastal terrain, such as that along the east-central and southern coasts of the United States, extends from inland scarps and terraces to the ocean (area C of the conceptual landscape, Figure 2). This terrain is characterized by (1) low scarps and terraces that were formed when the ocean was higher than at present; (2) streams, estuaries, and lagoons that are affected by tides; (3) ponds that are commonly associated with coastal sand dunes; and (4) barrier islands. Wetlands cover extensive areas in some coastal terrains (see Figure 18).

The interaction of ground water and surface water in coastal terrain is affected by discharge of ground water from regional flow systems and from local flow systems associated with scarps and terraces (Figure 23), evapotranspiration, and tidal flooding. The local flow systems associated with scarps and terraces are caused by the configuration of the water table near these features (see Box J). Where the water table has a downward break in slope near the top of scarps and terraces, downward components of ground-water flow are present; where the water table has an upward break in slope near the base of these features, upward components of ground-water flow are present.

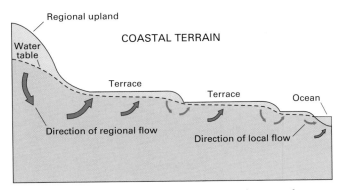

Figure 23. *In coastal terrain, small local ground-water flow cells associated with terraces overlie more regional ground-water flow systems. In the tidal zone, saline and brackish surface water mixes with fresh ground water from local and regional flow systems.*

Coastal terrain in Maryland. (Photograph by Robert Shedlock.)

Evapotranspiration directly from ground water is widespread in coastal terrain. The land surface is flat and the water table generally is close to land surface; therefore, many plants have root systems deep enough to transpire ground water at nearly the maximum potential rate. The result is that evapotranspiration causes a significant water loss, which affects the configuration of ground-water flow systems as well as how ground water interacts with surface water.

In the parts of coastal landscapes that are affected by tidal flooding, the interaction of ground water and surface water is similar to that in alluvial valleys affected by flooding. The principal difference between the two is that tidal flooding is more predictable in both timing and magnitude than river flooding. The other significant difference is in water chemistry. The water that moves into bank storage from rivers is generally fresh, but the water that moves into bank storage from tides generally is brackish or saline.

Estuaries are a highly dynamic interface between the continents and the ocean, where discharge of freshwater from large rivers mixes with saline water from the ocean. In addition, ground water discharges to estuaries and the ocean, delivering nutrients and contaminants directly to coastal waters. However, few estimates of the location and magnitude of ground-water discharge to coasts have been made.

In some estuaries, sulfate-rich regional ground water mixes with carbonate-rich local ground water and with chloride-rich seawater, creating sharp boundaries that separate plant and wildlife communities. Biological communities associated with these sharp boundaries are adapted to different hydrochemical conditions, and they undergo periodic stresses that result from inputs of water having different chemistry. The balance between river inflow and tides causes estuaries to retain much of the particulate and dissolved matter that is transported in surface and subsurface flows, including contaminants.

Tidal mangrove wetland in Florida. (Photograph by Virginia Carter.)

"Ground water discharges to estuaries and the ocean, delivering nutrients and contaminants directly to coastal waters"

Field Studies of Coastal Terrain

Along the Atlantic, Gulf of Mexico, and Arctic Coasts of the United States, broad coastal plains are transected by streams, scarps, and terraces. In some parts of these regions, local ground-water flow systems are associated with scarps and terraces, and freshwater wetlands commonly are present. Other parts of coastal regions are affected by tides, resulting in very complex flow and biogeochemical processes.

Underlying the broad coastal plain of the mid-Atlantic United States are sediments 600 or more feet thick. The sands and clays were deposited in stratigraphic layers that slope gently from west to east. Ground water moves regionally toward the east in the more permeable sand layers. These aquifers are separated by discontinuous layers of clay that restrict vertical ground-water movement. Near land surface, local ground-water flow systems are associated with changes in land slope, such as at major scarps and at streams.

Studies of the Dismal Swamp in Virginia and North Carolina provide examples of the interaction of ground water and wetlands near a coastal scarp. The Suffolk Scarp borders the west side of Great Dismal Swamp. Water-table wells and deeper piezometers placed across the scarp indicated a downward component of ground-water flow in the upland and an upward component of ground-water flow in the lowland at the edge of the swamp (Figure J–1A). However, at the edge of the swamp the direction of flow changed several times between May and October in 1982 because transpiration of ground water lowered the water table below the water level of the deep piezometer (Figure J–1B).

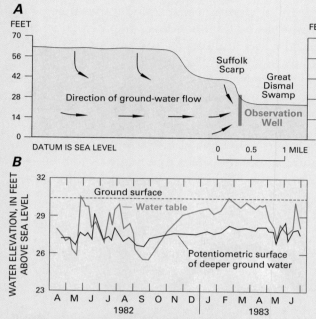

Figure J–1. Ground-water discharge at the edge of the Great Dismal Swamp in Virginia provides an example of local ground-water flow systems associated with coastal scarps (A). The vertical components of flow can change direction seasonally, partly because evapotranspiration discharges shallower ground water during part of the year (B). (Modified from Carter, Virginia, 1990, The Great Dismal Swamp—An illustrated case study, chapter 8, in Lugo, A.E., Brinson, Mark, and Brown, Sandra, eds., Ecosystems of the world, 15: Forested wetlands, Elsevier, Amsterdam, p. 201–211.) (Reprinted with permission from Elsevier Science-NL, Amsterdam, The Netherlands.)

Great Dismal Swamp, Virginia. (Photograph by Virginia Carter.)

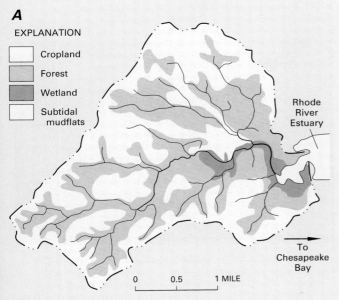

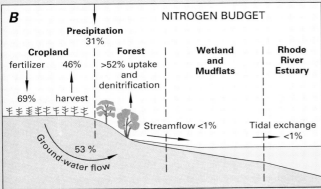

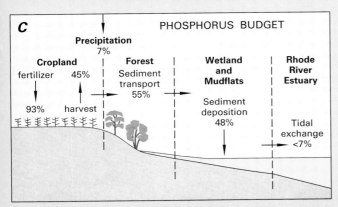

Figure J–2. Forests and wetlands separate cropland from streams in the Rhode River watershed in Maryland (A). More than half of the nitrogen applied to cropland is transported by ground water toward riparian forests and wetlands (B). More than half of the total phosphorus applied to cropland is transported by streams to wetlands and mudflats, where most is deposited in sediments (C). (Modified from Correll, D.L., Jordan, T.E., and Weller, D.E., 1992, Nutrient flux in a landscape—Effects of coastal land use and terrestrial community mosaic on nutrient transport to coastal waters: Estuaries, v. 15, no. 4, p. 431–442.) (Reprinted by permission of the Estuarine Research Federation.)

The gentle relief and sandy, well-drained soils of coastal terrain are ideal for agriculture. Movement of excess nutrients to estuaries are a particular problem in coastal areas because the slow rate of flushing of coastal bays and estuaries can cause them to retain nutrients. At high concentrations, nutrients can cause increased algal production, which results in overabundance of organic matter. This, in turn, can lead to reduction of dissolved oxygen in surface water to the extent that organisms are killed throughout large areas of estuaries and coastal bays.

Movement of nutrients from agricultural fields has been documented for the Rhode River watershed in Maryland (Figure J–2). Application of fertilizer accounts for 69 percent of nitrogen and 93 percent of phosphorus input to this watershed (Figure J–2B and J–2C). Almost all of the nitrogen that is not removed by harvested crops is transported in ground water and is taken up by trees in riparian forests and wetlands or is denitrified to nitrogen gas in ground water before it reaches streams. On the other hand, most of the phosphorus not removed by harvested crops is attached to soil particles and is transported only during heavy precipitation when sediment from fields is transported into streams and deposited in wetlands and subtidal mudflats at the head of the Rhode River estuary. Whether phosphorus is retained in sediments or is released to the water column depends in part on whether sediments are exposed to oxygen. Thus, the uptake of nutrients and their storage in riparian forests, wetlands, and subtidal mudflats in the Rhode River watershed has helped maintain relatively good water quality in the Rhode River estuary.

In other areas, however, agricultural runoff and input of nutrients have overwhelmed coastal systems, such as in the northern Gulf of Mexico near the mouth of the Mississippi River. The 1993 flood in the Mississippi River system delivered an enormous amount of nutrients to the Gulf of Mexico. Following the flood, oxygen-deficient sediments created areas of black sediment devoid of animal life in parts of the northern Gulf of Mexico.

Rhode River, Maryland. (Photograph by David Correll.)

GLACIAL AND DUNE TERRAIN

Glacial and dune terrain (area G of the conceptual landscape, Figure 2) is characterized by a landscape of hills and depressions. Although stream networks drain parts of these landscapes, many areas of glacial and dune terrain do not contribute runoff to an integrated surface drainage network. Instead, surface runoff from precipitation falling on the landscape accumulates in the depressions, commonly resulting in the presence of lakes and wetlands. Because of the lack of stream outlets, the water balance of these "closed" types of lakes and wetlands is controlled largely by exchange of water with the atmosphere (precipitation and evapotranspiration) and with ground water (see Box K).

Glacial terrain in Minnesota. (Photograph by Robert Karls.)

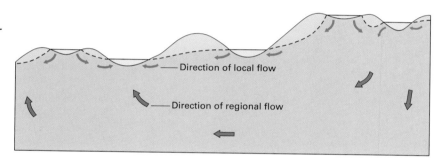

Figure 24. *In glacial and dune terrain, local, intermediate, and regional ground-water flow systems interact with lakes and wetlands. It is not uncommon for wetlands that recharge local ground-water flow systems to be present in lowlands and for wetlands that receive discharge from local ground water to be present in uplands.*

Lakes and wetlands in glacial and dune terrain can have inflow from ground water, outflow to ground water, or both (Figure 16). The interaction between lakes and wetlands and ground water is determined to a large extent by their position with respect to local and regional ground-water flow systems. A common conception is that lakes and wetlands that are present in topographically high areas recharge ground water, and that lakes and wetlands that are present in low areas receive discharge from ground water. However, lakes and wetlands underlain by deposits having low permeability can receive discharge from local ground-water flow systems even if they are located in a regional ground-water recharge area. Conversely, they can lose water to local ground-water flow systems even if they are located in a regional ground-water discharge area (Figure 24).

Lakes and wetlands in glacial and dune terrain underlain by highly permeable deposits commonly have ground-water seepage into one side and seepage to ground water on the other side. This relation is relatively stable because the water-table gradient between surface-water bodies in this type of setting is relatively constant. However, the boundary between inflow to the lake or wetland and outflow from it, termed the hinge line, can move up and down along the shoreline. Movement of the hinge line between inflow and outflow is a result of the changing slope of the water table in response to changes in ground-water recharge in the adjacent uplands.

Dune terrain in Nebraska. (Photograph by James Swinehart.)

Transpiration directly from ground water has a significant effect on the interaction of lakes and wetlands with ground water in glacial and dune terrain. Transpiration from ground water (Figure 7) has perhaps a greater effect on lakes and wetlands underlain by low-permeability deposits than in any other landscape. The lateral movement of ground water in low-permeability deposits may not be fast enough to supply the quantity of water at the rate it is removed by transpiration, resulting in deep and steep-sided cones of depression. These cones of depression commonly are present around the perimeter of the lakes and wetlands (Figure 7 and Box K).

In the north-central United States, cycles in the balance between precipitation and evapotranspiration that range from 5 to 30 years can result in large changes in water levels, chemical concentrations, and major-ion water type of individual wetlands. In some settings, repeated cycling of water between the surface and subsurface in the same locale results in evaporative concentration of solutes and eventually in mineral precipitation in the subsurface. In addition, these dynamic hydrological and chemical conditions can cause significant changes in the types, number, and distribution of wetland plants and invertebrate animals within wetlands. These changing hydrological conditions that range from seasons to decades are an essential process for rejuvenating wetlands that provide ideal habitat and feeding conditions for migratory waterfowl.

"The hydrological and chemical characteristics of lakes and wetlands in glacial and dune terrain are determined to a large extent by their position with respect to local and regional ground-water flow systems"

Field Studies of Glacial and Dune Terrain

Glacial terrain and dune terrain are characterized by land-surface depressions, many of which contain lakes and wetlands. Although much of the glacial terrain covering the north-central United States (see index map) has low topographic relief, neighboring lakes and wetlands are present at a sufficiently wide range of altitudes to result in many variations in how they interact with ground water, as evidenced by the following examples.

The Cottonwood Lake area, near Jamestown, North Dakota, is within the prairie-pothole region of North America. The hydrologic functions of these small depressional wetlands are highly variable in space and time. With respect to spatial variation, some wetlands recharge ground water, some receive ground-water inflow and have outflow to ground water, and some receive ground-water discharge. Wetland P1 provides an example of how their functions can vary in time. The wetland receives ground-water discharge most of the time; however, transpiration of ground water by plants around the perimeter of the wetland can cause water to seep from the wetland. Seepage from wetlands commonly is assumed to be ground-water recharge, but in cases like Wetland P1, the water is actually lost to transpiration. This process results in depressions in the water table around the perimeter of the wetland at certain times, as shown in

Cottonwood Lake area, North Dakota. (Photograph by Thomas Winter.)

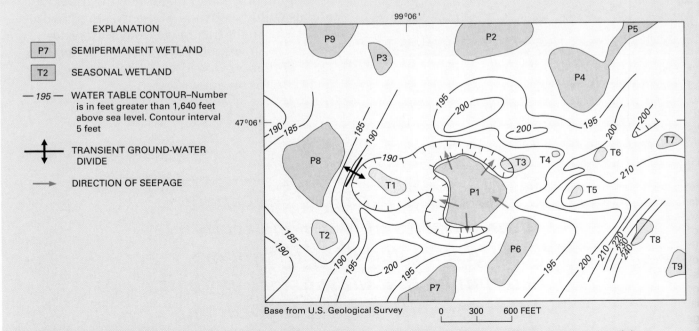

Figure K–1. Transpiration directly from ground water causes cones of depression to form by late summer around the perimeter of prairie pothole Wetland P1 in the Cottonwood Lake area in North Dakota. (Modified from Winter, T.C., and Rosenberry, D.O., 1995, The interaction of ground water with prairie pothole wetlands in the Cottonwood Lake area, east-central North Dakota, 1979–1990: Wetlands, v. 15, no. 3, p. 193–211.) (Used with permission.)

Figure K–1. Transpiration-induced depressions in the water table commonly are filled in by recharge during the following spring, but then form again to some extent by late summer nearly every year.

Nevins Lake, a closed lake in the Upper Peninsula of Michigan, illustrates yet another type of interaction of lakes with ground water in glacial terrain. Water-chemistry studies of Nevins Lake indicated that solutes such as calcium provide an indicator of ground-water inflow to the lake. Immediately following spring snowmelt, the mass of dissolved calcium in the lake increased rapidly because of increased ground-water inflow. Calcium then decreased steadily throughout the summer and early fall as the lake received less ground-water inflow (Figure K–2). This pattern varied annually depending on the amount of ground-water recharge from snowmelt and spring rains. The chemistry of water in the pores of the lake sediments was used to determine the spatial variability in the direction of seepage on the side of the lake that had the most ground-water inflow. Seepage was always out of the lake at the sampling site farthest from shore and was always upward into the lake at the site nearest to shore. Flow reversals were documented at sites located at intermediate distances from shore.

Nevins Lake, Michigan. (Photograph by David Krabbenhoft.)

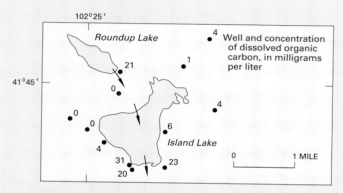

Figure K–3. Seepage from lakes in the sandhills of Nebraska causes plumes of dissolved organic carbon to be present in ground water on the downgradient sides of the lakes. (Modified from LaBaugh, J.W., 1986, Limnological characteristics of selected lakes in the Nebraska sandhills, U.S.A., and their relation to chemical characteristics of adjacent ground water: Journal of Hydrology, v. 86, p. 279–298.) (Reprinted with permission of Elsevier Science-NL, Amsterdam, The Netherlands.)

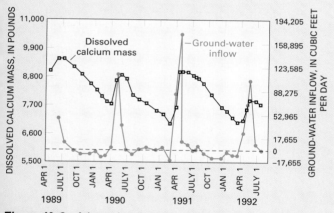

Figure K–2. A large input of ground water during spring supplies the annual input of calcium to Nevins Lake in the Upper Peninsula of Michigan. (Modified from Krabbenhoft, D.P., and Webster, K.E., 1995, Transient hydrogeological controls on the chemistry of a seepage lake: Water Resources Research, v. 31, no. 9, p. 2295–2305.)

Dune terrain also commonly contains lakes and wetlands. Much of the central part of western Nebraska, for example, is covered by sand dunes that have lakes and wetlands in most of the lowlands between the dunes. Studies of the interaction of lakes and wetlands with ground water at the Crescent Lake National Wildlife Refuge indicate that most of these lakes have seepage inflow from ground water and seepage outflow to ground water. The chemistry of inflowing ground water commonly has an effect on lake water chemistry. However, the chemistry of lake water can also affect ground water in areas of seepage from lakes. In the Crescent Lake area, for example, plumes of lake water were detected in ground water downgradient from the lakes, as indicated by the plume of dissolved organic carbon downgradient from Roundup Lake and Island Lake (Figure K–3).

Island Lake, Nebraska. (Photograph by Thomas Winter.)

KARST TERRAIN

Karst may be broadly defined as all landforms that are produced primarily by the dissolution of rocks, mainly limestone and dolomite. Karst terrains (area K of the conceptual landscape, Figure 2) are characterized by (1) closed surface depressions of various sizes and shapes known as sinkholes, (2) an underground drainage network that consists of solution openings that range in size from enlarged cracks in the rock to large caves, and (3) highly disrupted surface drainage systems, which relate directly to the unique character of the underground drainage system.

Dissolution of limestone and dolomite guides the initial development of fractures into solution holes that are diagnostic of karst terrain. Perhaps nowhere else is the complex interplay between hydrology and chemistry so important to changes in landform. Limestone and dolomite weather quickly, producing calcium and magnesium carbonate waters that are relatively high in ionic strength. The increasing size of solution holes allows higher ground-water flow rates across a greater surface area of exposed minerals, which stimulates the dissolution process further, eventually leading to development of caves. Development of karst terrain also involves biological processes. Microbial production of carbon dioxide in the soil affects the carbonate equilibrium of

Big Spring, Missouri. (Photograph by James Barks.)

water as it recharges ground water, which then affects how much mineral dissolution will take place before solute equilibrium is reached.

Ground-water recharge is very efficient in karst terrain because precipitation readily infiltrates through the rock openings that intersect the land surface. Water moves at greatly different rates through karst aquifers; it moves slowly through fine fractures and pores and rapidly through solution-enlarged fractures and conduits. As a result, the water discharging from many springs in karst terrain may be a combination of relatively slow-moving water draining from pores and rapidly moving storm-derived water. The slow-moving component tends to reflect the chemistry of the aquifer materials, and the more rapidly moving water associated with recent rainfall tends to reflect the chemical characteristics of precipitation and surface runoff.

Water movement in karst terrain is especially unpredictable because of the many paths ground water takes through the maze of fractures and solution openings in the rock (see Box L). Because of the large size of interconnected openings in well-developed karst systems, karst terrain can have true underground streams. These underground streams can have high rates of flow, in some places as great as rates of flow in surface streams. Furthermore, it is not unusual for medium-sized streams to disappear into the rock openings, thereby completely disrupting

the surface drainage system, and to reappear at the surface at another place. Seeps and springs of all sizes are characteristic features of karst terrains. Springs having sufficiently large ground-water recharge areas commonly are the source of small- to medium-sized streams and constitute a large

Stream disappearing into sinkhole in karst terrain in Texas. (Photograph by Jon Gilhousen.)

part of tributary flow to larger streams. In addition, the location where the streams emerge can change, depending on the spatial distribution of ground-water recharge in relation to individual precipitation events. Large spring inflows to streams in karst terrain contrast sharply with the generally more diffuse ground-water inflow characteristic of streams flowing across sand and gravel aquifers.

Because of the complex patterns of surface-water and ground-water flow in karst terrain, many studies have shown that surface-water drainage divides and ground-water drainage divides do not coincide. An extreme example is a stream that disappears in one surface-water basin and reappears in another basin. This situation complicates the identification of source areas for water and associated dissolved constituents, including contaminants, in karst terrain.

Water chemistry is widely used for studying the hydrology of karst aquifers. Extensive tracer studies (see Box G) and field mapping to locate points of recharge and discharge have been used to estimate the recharge areas of springs, rates of ground-water movement, and the water balance of aquifers. Variations in parameters such as temperature, hardness, calcium/magnesium ratios, and other chemical characteristics have been used to identify areas of ground-water recharge, differentiate rapid- and slow-moving ground-water flow paths, and compare springflow characteristics in different regions. Rapid transport of contaminants within karst aquifers and to springs has been documented in many locations. Because of the rapid movement of water in karst aquifers, water-quality problems that might be localized in other aquifer systems can become regional problems in karst systems.

Some landscapes considered to be karst terrain do not have carbonate rocks at the land surface. For example, in some areas of the southeastern United States, surficial deposits overlie carbonate rocks, resulting in a "mantled" karst terrain. Lakes and wetlands in mantled karst terrain interact with shallow ground water in a manner similar to that in sandy glacial and dune terrains. The difference between how lakes and wetlands interact with ground water in sandy glacial and dune terrain and how they interact in the mantled karst is related to the buried carbonate rocks. If dissolution of the buried carbonate rocks causes slumpage of an overlying confining bed, such that water can move freely through the confining bed, the lakes and wetlands also can be affected by changing hydraulic heads in the aquifers underlying the confining bed (see Box L).

Field Studies of Karst Terrain

Karst terrain is characteristic of regions that are underlain by limestone and dolomite bedrock. In many karst areas, the carbonate bedrock is present at land surface, but in other areas it may be covered by other deposits and is referred to as "mantled" karst. The Edwards Aquifer in south-central Texas is an example of karst terrain where the limestones and dolomites are exposed at land surface (Figure L–1). In this outcrop area, numerous solution cavities along vertical joints and sinkholes provide an efficient link between the land surface and the water table. Precipitation on the outcrop area tends to infiltrate rapidly into the ground, recharging ground water. In addition, a considerable amount of recharge to the aquifer is provided by losing streams that cross the outcrop area. Even the largest streams that originate to the north are dry in the outcrop area for most of the year. The unusual highway signs in this area go beyond local pride in a prolific water supply—they reflect a clear understanding of how vulnerable this water supply is to contamination by human activities at the land surface.

Just as solution cavities are major avenues for ground-water recharge, they also are focal points for ground-water discharge from karst aquifers. For example, springs near the margin of the Edwards Aquifer provide a continuous source of water for streams to the south.

An example of mantled karst can be found in north-central Florida, a region that has many sinkhole lakes. In this region, unconsolidated deposits overlie the highly soluble limestone of the Upper Floridan aquifer. Most land-surface depressions containing lakes in Florida are formed when unconsolidated surficial deposits slump into sinkholes that form in the underlying limestone. Thus, although the lakes are not situated directly in limestone, the sinkholes in the bedrock underlying lakes commonly have a significant effect on the hydrology of the lakes.

Edwards Aquifer recharge area, Texas. (Photograph by Rene Barker.)

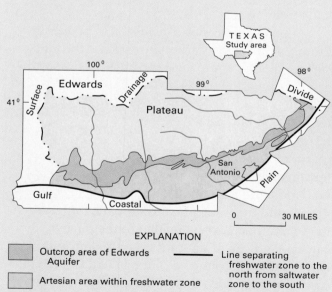

Figure L–1. *A large area of karst terrain is associated with the Edwards Aquifer in south-central Texas. Large streams lose a considerable amount of water to ground water as they traverse the outcrop area of the Edwards Aquifer. (Modified from Brown, D.S., and Patton, J.T., 1995, Recharge to and discharge from the Edwards Aquifer in the San Antonio area, Texas, 1995: U.S. Geological Survey Open-File Report 96–181, 2 p.)*

Lake Barco is one of numerous lakes occupying depressions in northern Florida. Results of a study of the interaction of Lake Barco with ground water indicated that shallow ground water flows into the northern and northeastern parts of the lake, and lake water seeps out to shallow ground water in the western and southern parts (Figure L–2A). In addition, ground-water flow is downward beneath most of Lake Barco (Figure L–2B).

The studies of lake and ground-water chemistry included the use of tritium, chlorofluorocarbons (CFCs), and isotopes of oxygen (see Box G). The results indicated significant differences in the chemistry of (1) shallow ground water flowing into Lake Barco, (2) Lake Barco water, (3) shallow ground water downgradient from Lake Barco, and (4) deeper ground water beneath Lake Barco. Oxygen-rich lake water moving through the organic-rich lake sediments is reduced, resulting in discharge of oxygen-depleted water into the ground water beneath Lake Barco. This downward-moving ground water may have an undesired effect on the chemical quality of ground water in the underlying Upper Floridan aquifer, which is the principal source of water supply for the region. The patterns of ground-water movement determined from hydraulic-head data were corroborated by chemical tracers. For example, the dates that ground water in different parts of the flow system was recharged, as determined from CFC dating, show a fairly consistent increase in the length of time since recharge with depth (Figure L–2C).

Lake Barco, Florida. (Photograph by Terrie Lee.)

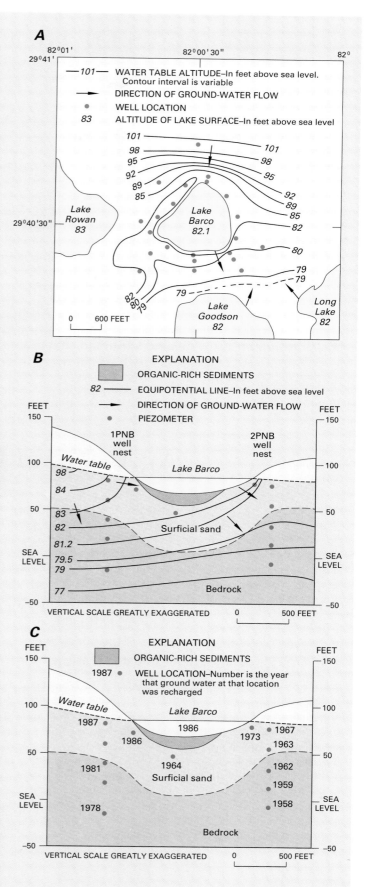

Figure L–2. *Lake Barco, in northern Florida, is a flow-through lake with respect to ground water (A and B). The dates that ground water in different parts of the ground-water system was recharged indicate how long it takes water to move from the lake or water table to a given depth (C). (Modified from Katz, B.G., Lee, T.M., Plummer, L.N., and Busenberg, E., 1995, Chemical evolution of groundwater near a sinkhole lake, northern Florida, 1. Flow patterns, age of groundwater, and influence of lake water leakage: Water Resources Research, v. 31, no. 6, p. 1549–1564.)*

EFFECTS OF HUMAN ACTIVITIES ON THE INTERACTION OF GROUND WATER AND SURFACE WATER

Human activities commonly affect the distribution, quantity, and chemical quality of water resources. The range in human activities that affect the interaction of ground water and surface water is broad. The following discussion does not provide an exhaustive survey of all human effects but emphasizes those that are relatively widespread. To provide an indication of the extent to which humans affect the water resources of virtually all landscapes, some of the most relevant structures and features related to human activities are superimposed on various parts of the conceptual landscape (Figure 25).

The effects of human activities on the quantity and quality of water resources are felt over a wide range of space and time scales. In the following discussion, "short term" implies time scales from hours to a few weeks or months, and "long term" may range from years to decades. "Local scale" implies distances from a few feet to a few thousand feet and areas as large as a few square miles, and "subregional and regional scales" range from tens to thousands of square miles. The terms point source and nonpoint source with respect to discussions of contamination are used often; therefore, a brief discussion of the meaning of these terms is presented in Box M.

Agricultural Development

Agriculture has been the cause of significant modification of landscapes throughout the world. Tillage of land changes the infiltration and runoff characteristics of the land surface, which affects recharge to ground water, delivery of water and sediment to surface-water bodies, and evapotranspiration. All of these processes either directly or indirectly affect the interaction of ground water and surface water. Agriculturalists are aware of the substantial negative effects of agriculture on water resources and have developed methods to alleviate some of these effects. For example, tillage practices have been modified to maximize retention of water in soils and to minimize erosion of soil from the land into surface-water bodies. Two activities related to agriculture that are particularly relevant to the interaction of ground water and surface water are irrigation and application of chemicals to cropland.

Figure 25. Human activities and structures, as depicted by the distribution of various examples in the conceptual landscape, affect the interaction of ground water and surface water in all types of landscapes.

Point and Nonpoint Sources of Contaminants

Contaminants may be present in water or in air as a result of natural processes or through mechanisms of displacement and dispersal related to human activities. Contaminants from point sources discharge either into ground water or surface water through an area that is small relative to the area or volume of the receiving water body. Examples of point sources include discharge from sewage-treatment plants, leakage from gasoline storage tanks, and seepage from landfills (Figure M–1).

Nonpoint sources of contaminants introduce contaminants to the environment across areas that are large compared to point sources, or nonpoint sources may consist of multiple, closely spaced point sources. A nonpoint source of contamination that can be present anywhere, and affect large areas, is deposition from the atmosphere, both by precipitation (wet deposition) or by dry fallout (dry deposition). Agricultural fields, in aggregate, represent large areas through which fertilizers and pesticides can be released to the environment.

The differentiation between point and nonpoint sources of contamination is arbitrary to some extent and may depend in part on the scale at which a problem is considered. For example, emissions from a single smokestack is a point source, but these emissions may be meaningless in a regional analysis of air pollution. However, a fairly even distribution of tens or hundreds of smokestacks might be considered as a nonpoint source. As another example, houses in suburban areas that do not have a combined sewer system have individual septic tanks. At the local scale, each septic tank may be considered as point source of contamination to shallow ground water. At the regional scale, however, the combined contamination of ground water from all the septic tanks in a suburban area may be considered a nonpoint source of contamination to a surface-water body.

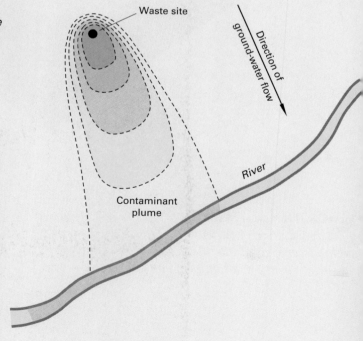

Figure M–1. The transport of contamination from a point source by ground water can cause contamination of surface water, as well as extensive contamination of ground water.

IRRIGATION SYSTEMS

Surface-water irrigation systems represent some of the largest integrated engineering works undertaken by humans. The number of these systems greatly increased in the western United States in the late 1840s. In addition to dams on streams, surface-water irrigation systems include (1) a complex network of canals of varying size and carrying capacity that transport water, in many cases for a considerable distance, from a surface-water source to individual fields, and (2) a drainage system to carry away water not used by plants that may be as extensive and complex as the supply system. The drainage system may include underground tile drains. Many irrigation systems that initially used only surface water now also use ground water. The pumped ground water commonly is used directly as irrigation water, but in some cases the water is distributed through the system of canals.

Average quantities of applied water range from several inches to 20 or more inches of water per year, depending on local conditions, over the entire area of crops. In many irrigated areas, about 75 to 85 percent of the applied water is lost to evapotranspiration and retained in the crops (referred to as consumptive use). The remainder of the water either infiltrates through the soil zone to recharge ground water or it returns to a local surface-water body through the drainage system (referred to as irrigation return flow). The quantity of irrigation water that recharges ground water usually is large relative to recharge from precipitation because large irrigation systems commonly are in regions of low precipitation and low natural recharge. As a result, this large volume of artificial recharge can cause the water table to rise (see Box N), possibly reaching the land surface in some areas and waterlogging the fields. For this reason, drainage systems that maintain the level of the water table below the root zone of the crops, generally 4 to 5 feet below the land surface, are an essential component of some irrigation systems. The permanent rise in the water table that is maintained by continued recharge from irrigation return flow commonly results in an increased outflow of shallow ground water to surface-water bodies downgradient from the irrigated area.

Gravity irrigation using surface water in Nebraska. (Photograph by Les Sheffield.)

Effects of Irrigation Development on the Interaction of Ground Water and Surface Water

Nebraska ranks second among the States with respect to the area of irrigated acreage and the quantity of water used for irrigation. The irrigation water is derived from extensive supply systems that use both surface water and ground water (Figure N–1). Hydrologic conditions in different parts of Nebraska provide a number of examples of the broad-scale effects of irrigation development on the interactions of ground water and surface water. As would be expected, irrigation systems based on surface water are always located near streams. In general, these streams are perennial and (or) have significant flow for at least part of the year. In contrast, irrigation systems based on ground water can be located nearly anywhere that has an adequate ground-water resource. Areas of significant rise and decline in ground-water levels due to irrigation systems are shown in Figure N–2. Ground-water levels rise in some areas irrigated with surface water and decline in some areas irrigated with ground water. Rises in ground-water levels near streams result in increased ground-water inflow to gaining streams or decreased flow from the stream to ground water for losing streams. In some areas, it is possible that a stream that was losing water before development of irrigation could become a gaining stream following irrigation. This effect of surface-water irrigation probably caused the rises in ground-water levels in areas F and G in south-central Nebraska (Figure N–2).

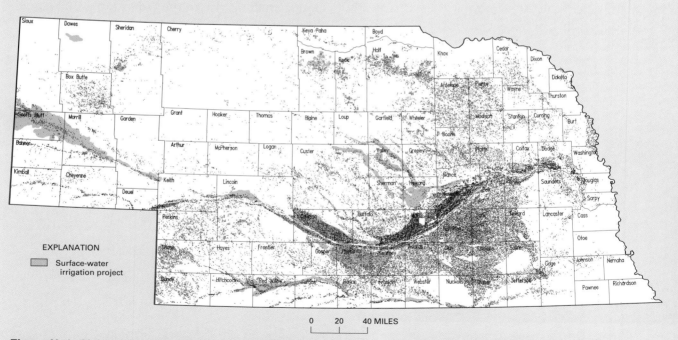

Figure N–1. *Nebraska is one of the most extensively irrigated States in the Nation. The irrigation water comes from both ground-water and surface-water sources. Dots are irrigation wells. (Map provided by the University of Nebraska, Conservation and Survey Division.)*

...and irrigated by using ground-water center-pivot sprinklers in Nebraska. (Photograph by Les Sheffield.)

Average annual precipitation ranges from less than 15 inches in western Nebraska to more than 30 inches in eastern Nebraska. A large concentration of irrigation wells is present in area E (Figure N–2). The ground-water withdrawals by these wells caused declines in ground-water levels that could not be offset by recharge from precipitation and the presence of nearby flowing streams. In this area, the withdrawals cause decreases in ground-water discharge to the streams and (or) induce flow from the streams to shallow ground water. In contrast, the density of irrigation wells in areas A, B, and C is less than in area E, but water-level declines in these three western areas are similar to area E. The similar decline caused by fewer wells in the west compared to the east is related to less precipitation, less ground-water recharge, and less streamflow available for seepage to ground water.

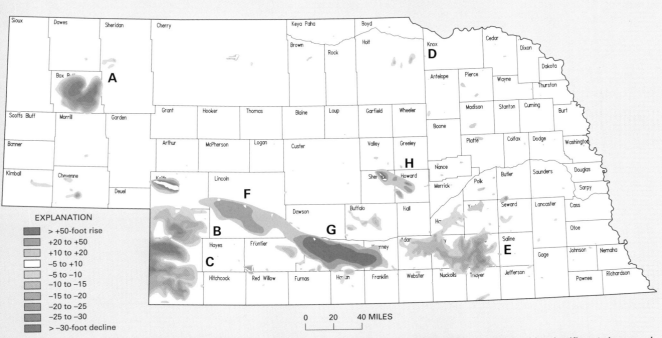

Figure N–2. The use of both ground water and surface water for irrigation in Nebraska has resulted in significant rises and declines of ground-water levels in different parts of the State. (Map provided by the University of Nebraska, Conservation and Survey Division.)

Although early irrigation systems made use of surface water, the development of large-scale sprinkler systems in recent decades has greatly increased the use of ground water for irrigation for several reasons: (1) A system of supply canals is not needed, (2) ground water may be more readily available than surface water, and (3) many types of sprinkler systems can be used on irregular land surfaces; the fields do not have to be as flat as they do for gravity-flow, surface-water irrigation.

Sprinkler irrigation using ground water in Nebraska. (Photograph by Les Sheffield.)

Whether ground water or surface water was used first to irrigate land, it was not long before water managers recognized that development of either water resource could affect the other. This is particularly true in many alluvial aquifers in arid regions where much of the irrigated land is in valleys.

Significant changes in water quality accompany the movement of water through agricultural fields. The water lost to evapotranspiration is relatively pure; therefore, the chemicals that are left behind precipitate as salts and accumulate in the soil zone. These continue to increase as irrigation continues, resulting in the dissolved-solids concentration in the irrigation return flows being significantly higher in some areas than that in the original irrigation water. To prevent excessive buildup of salts in the soil, irrigation water in excess of the needs of the crops is required to dissolve and flush out the salts and transport them to the ground-water system. Where these dissolved solids reach high concentrations, the artificial recharge from irrigation return flow can result in degradation of the quality of ground water and, ultimately, the surface water into which the ground water discharges.

"Whether ground water or surface water was used first to irrigate land, it was not long before water managers recognized that development of either water resource could affect the other"

USE OF AGRICULTURAL CHEMICALS

Applications of pesticides and fertilizers to cropland can result in significant additions of contaminants to water resources. Some pesticides are only slightly soluble in water and may attach (sorb) to soil particles instead of remaining in solution; these compounds are less likely to cause contamination of ground water. Other pesticides, however, are detected in low, but significant, concentrations in both ground water and surface water. Ammonium, a major component of fertilizer and manure, is very soluble in water, and increased concentrations of nitrate that result from nitrification of ammonium commonly are present in both ground water and surface water associated with agricultural lands (see Box O). In addition to these nonpoint sources of water contamination, point sources of contamination are common in agricultural areas where livestock are concentrated in small areas, such as feedlots. Whether the initial contamination is present in ground water or surface water is somewhat immaterial because the close interaction of the two sometimes results in both being contaminated (see Box P).

Applying chemicals to cropland in Maryland. (Photograph by David Usher.)

"Whether the initial contamination is present in ground water or surface water is somewhat immaterial because the close interaction of the two sometimes results in both being contaminated"

Effects of Nitrogen Use on the Quality of Ground Water and Surface Water

Nitrate contamination of ground water and surface water in the United States is widespread because nitrate is very mobile in the environment. Nitrate concentrations are increasing in much of the Nation's water, but they are particularly high in ground water in the midcontinent region of the United States. Two principal chemical reactions are important to the fate of nitrogen in water: (1) fertilizer ammonium can be nitrified to form nitrate, which is very mobile as a dissolved constituent in shallow ground water, and (2) nitrate can be denitrified to produce nitrogen gas in the presence of chemically reducing conditions if a source of dissolved organic carbon is available.

High concentrations of nitrate can contribute to excessive growth of aquatic plants, depletion of oxygen, fishkills, and general degradation of aquatic habitats. For example, a study of Waquoit Bay in Massachusetts linked the decline in eelgrass beds since 1950 to a progressive increase in nitrate input due to expansion of domestic septic-field developments in the drainage basin (Figure O–1). Loss of eelgrass is a concern because this aquatic plant stabilizes sediment and provides ideal habitat for juvenile fish and other fauna in coastal bays and estuaries. Larger nitrate concentrations supported algal growth that caused turbidity and shading, which contributed to the decline of eelgrass.

Waquoit Bay, Massachusetts. (Photograph by Ivan Valiela.)

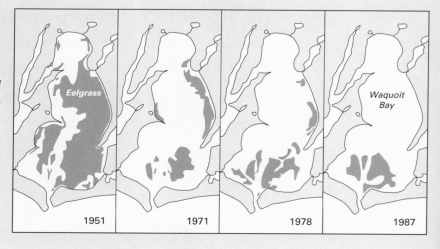

Figure O–1. The areal extent of eelgrass in Waquoit Bay, Massachusetts, decreased markedly between 1951 and 1987 because of increased inputs of nitrogen related to domestic septic-field developments. (Modified from Valiela, I., Foreman, K., LaMontagne, M., Hersh, D., Costa, J., Peckol, P., DeMeo-Andeson, B., D'Avanzo, C., Babione, M., Sham, C.H., Brawley, J., and Lajtha, K., 1992, Couplings of watersheds and coastal waters—Sources and consequences of nutrient enrichment in Waquoit Bay, Massachusetts: Estuaries, v. 15, no. 4, p. 433–457.) (Reprinted by permission of the Estuarine Research Federation.)

Significant denitrification has been found to take place at locations where oxygen is absent or present at very low concentrations and where suitable electron-donor compounds, such as organic carbon, are available. Such locations include the interface of aquifers with silt and clay confining beds and along riparian zones adjacent to streams. For example, in a study on the eastern shore of Maryland, nitrogen isotopes and other environmental tracers were used to show that the degree of denitrification that took place depended on the extent of interaction between ground-water and the chemically reducing sediments near or below the bottom of the Aquia Formation. Two drainage basins were studied: Morgan Creek and Chesterville Branch (Figure O–2). Ground-water discharging beneath both streams had similar nitrate concentration when recharged. Significant denitrification took place in the Morgan Creek basin where a large fraction of local ground-water flow passed through the reducing sediments, which are present at shallow depths (3 to 10 feet) in this area. Evidence for the denitrification included decreases in nitrate concentrations along the flow path to Morgan Creek and enrichment of the ^{15}N isotope. Much less denitrification took place in the Chesterville Branch basin because the top of the reducing sediments are deeper (10 to 20 feet) in this area and a smaller fraction of ground-water flow passed through those sediments.

River in Delmarva Peninsula, Maryland. (Photograph by Pixie Hamilton.)

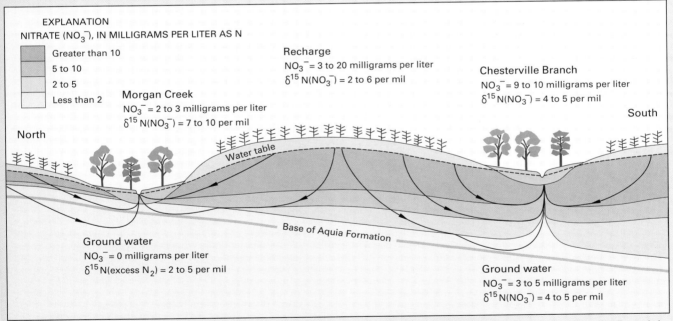

Figure O–2. Denitrification had a greater effect on ground water discharging to Morgan Creek than to Chesterville Branch in Maryland because a larger fraction of the local flow system discharging to Morgan Creek penetrated the reduced calcareous sediments near or below the bottom of the Aquia Formation than the flow system associated with the Chesterville Branch. (Modified from Bolke, J.K., and Denver, J.M., 1995, Combined use of ground-water dating, chemical, and isotopic analyses to resolve the history and fate of nitrate contamination in two agricultural watersheds, Atlantic coastal plain, Maryland: Water Resources Research, v. 31, no. 9, p. 2319–2337.)

Effects of Pesticide Application to Agricultural Lands on the Quality of Ground Water and Surface Water

Pesticide contamination of ground water and surface water has become a major environmental issue. Recent studies indicate that pesticides applied to cropland can contaminate the underlying ground water and then move along ground-water flow paths to surface water. In addition, as indicated by the following examples, movement of these pesticides between surface water and ground water can be dynamic in response to factors such as bank storage during periods of high runoff and ground-water withdrawals.

A study of the sources of atrazine, a widely used herbicide detected in the Cedar River and its associated alluvial aquifer in Iowa, indicated that ground water was the major source of atrazine in the river during base-flow conditions. In addition, during periods of high streamflow, surface water containing high concentrations of atrazine moved into the bank sediments and alluvial aquifer, then slowly discharged back to the river as the river level declined. Reversals of flow related to bank storage were documented using data for three sampling periods (Figure P–1). The first sampling (Figure P–1A) was before atrazine was applied to cropland, when concentrations in the river and aquifer were relatively low. The second sampling (Figure P–1B) was after atrazine was applied to cropland upstream. High streamflow at this time caused the river stage to peak almost 6 feet above its base-flow level, which caused the herbicide to move with the river water into the aquifer. By the third sampling date (Figure P–1C), the hydraulic gradient between the river and the alluvial aquifer had reversed again, and atrazine-contaminated water discharged back into the river.

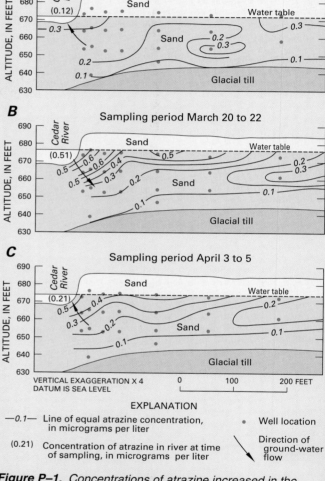

Cedar River, Iowa. (Photograph by Douglas Schnoebelen.)

Figure P–1. Concentrations of atrazine increased in the Cedar River in Iowa following applications of the chemical on agricultural areas upstream from a study site. During high streamflow (B), the contaminated river water moved into the alluvial aquifer as bank storage, contaminating ground water. After the river level declined (C), part of the contaminated ground water returned to the river. (Modified from Squillace, P.J., Thurman, E.M., and Furlong, E.T., 1993, Groundwater as a nonpoint source of atrazine and deethylatrazine in a river during base flow conditions: Water Resources Research, v. 29, no. 6, p. 1719–1729.)

In a second study, atrazine was detected in ground water in the alluvial aquifer along the Platte River near Lincoln, Nebraska. Atrazine is not applied in the vicinity of the well field, so it was suspected that ground-water withdrawals at the well field caused contaminated river water to move into the aquifer. To define the source of the atrazine, water samples were collected from monitoring wells located at different distances from the river near the well field. The pattern of concentrations of atrazine in the ground water indicated that peak concentrations of the herbicide showed up sooner in wells close to the river compared to wells farther away (Figure P–2). Peak concentrations of atrazine in ground water were much higher and more distinct during periods of large ground-water withdrawals (July and August) than during periods of much smaller withdrawals (May to early June).

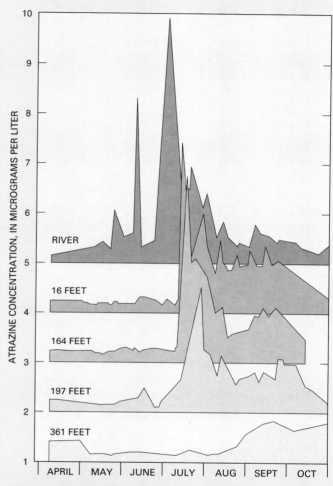

Figure P–2. Pumping of municipal water-supply wells near Lincoln, Nebraska, has induced Platte River water contaminated with atrazine to flow into the aquifer. Distances shown are from river to monitoring well. (Modified from Duncan, D., Pederson, D.T., Shepherd, T.R., and Carr, J.D., 1991, Atrazine used as a tracer of induced recharge: Ground Water Monitoring Review, v. 11, no. 4, p. 144–150.) (Used with permission.)

Platte River near Lincoln, Nebraska. (Photograph by Ralph Davis.)

Urban and Industrial Development

Point sources of contamination to surface-water bodies are an expected side effect of urban development. Examples of point sources include direct discharges from sewage-treatment plants, industrial facilities, and stormwater drains. These facilities and structures commonly add sufficient loads of a variety of contaminants to streams to strongly affect the quality of the stream for long distances downstream. Depending on relative flow magnitudes of the point source and of the stream, discharge from a point source such as a sewage-treatment plant may represent a large percentage of the water in the stream directly downstream from the source. Contaminants in streams can easily affect ground-water quality, especially where streams normally seep to ground water, where ground-water withdrawals induce seepage from the stream, and where floods cause stream water to become bank storage.

Point sources of contamination to ground water can include septic tanks, fluid storage tanks, landfills, and industrial lagoons. If a contaminant is soluble in water and reaches the water table, the contaminant will be transported by the slowly moving ground water. If the source continues to supply the contaminant over a period of time, the distribution of the dissolved contaminant will take a characteristic "plumelike" shape (see Box M). These contaminant plumes commonly discharge into a nearby surface-water body. If the concentration of contaminant is low and the rate of discharge of plume water also is small relative to the volume of the receiving surface-water body, the discharging contaminant plume will have only a small, or perhaps unmeasurable, effect on the quality of the receiving surface-water body. Furthermore, biogeochemical processes may decrease the concentration of the contaminant as it is transported through the shallow ground-water system and the hyporheic zone. On the other hand, if the discharge of the contaminant plume is large or has high concentrations of contaminant, it could significantly affect the quality of the receiving surface-water body.

Point source of urban runoff to surface water in Indiana. (Photograph by Charles Crawford.)

"Contaminants in streams can easily affect ground-water quality, especially where streams normally seep to ground water, where ground-water withdrawals induce seepage from the stream, and where floods cause stream water to become bank storage"

Drainage of the Land Surface

In landscapes that are relatively flat, have water ponded on the land surface, or have a shallow water table, drainage of land is a common practice preceding agricultural and urban development. Drainage can be accomplished by constructing open ditches or by burying tile drains beneath the land surface. In some glacial terrain underlain by deposits having low permeability, drainage of lakes and wetlands can change the areal distribution of ground-water recharge and discharge, which in turn can result in significant changes in the biota that are present and in the chemical and biological processes that take place in wetlands. Furthermore, these changes can ultimately affect the baseflow to streams, which in turn affects riverine ecosystems. Drainage also alters the water-holding capacity of topographic depressions as well as the surface runoff rates from land having very low slopes. More efficient runoff caused by drainage systems results in decreased recharge to ground water and greater contribution to flooding.

Drainage of the land surface is common in regions having extensive wetlands, such as coastal, riverine, and some glacial-lake landscapes. Construction of artificial drainage systems is extensive in these regions because wetland conditions generally result in deep, rich, organic soils that are much prized for agriculture. In the most extensive artificially drained part of the Nation, the glacial terrain of the upper Midwest, it is estimated that more than 50 percent of the original wetland areas have been destroyed. In Iowa alone, the destruction exceeds 90 percent. Although some wetlands were destroyed by filling, most were destroyed by drainage.

Artificial drainage in Minnesota. (Photograph by David Lorenz.)

Modifications to River Valleys

CONSTRUCTION OF LEVEES

Levees are built along riverbanks to protect adjacent lands from flooding. These structures commonly are very effective in containing smaller magnitude floods that are likely to occur regularly from year to year. Large floods that occur much less frequently, however, sometimes overtop or breach the levees, resulting in widespread flooding. Flooding of low-lying land is, in a sense, the most visible and extreme example of the interaction of ground water and surface water. During flooding, recharge to ground water is continuous; given sufficient time, the water table may rise to the land surface and completely saturate the shallow aquifer (see Figure 12). Under these conditions, an extended period of drainage from the shallow aquifer takes place after the floodwaters recede. The irony of levees as a flood protection mechanism is that if levees fail during a major flood, the area, depth, and duration of flooding in some areas may be greater than if levees were not present.

Breached levee along the Mississippi River. (Photograph courtesy of the St. Louis Post Dispatch.)

CONSTRUCTION OF RESERVOIRS

The primary purpose of reservoirs is to store water for uses such as public water supply, irrigation, flood attenuation, and generation of electric power. Reservoirs also can provide opportunities for recreation and wildlife habitat. Water needs to be stored in reservoirs because streamflow is highly variable, and the times when streamflow is abundant do not necessarily coincide with the times when the water is needed. Streamflow can vary daily in response to individual storms and seasonally in response to variation in weather patterns.

The effects of reservoirs on the interaction of ground water and surface water are greatest near the reservoir and directly downstream from it. Reservoirs can cause a permanent rise in the water table that may extend a considerable distance from the reservoir, because the base level of the stream, to which the ground-water gradients had adjusted, is raised to the higher reservoir levels. Near the dam, reservoirs commonly lose water to shallow ground water, but this water

Reservoir in California. (Photograph by Michael Collier.)

commonly returns to the river as base flow directly downstream from the dam. In addition, reservoirs can cause temporary bank storage at times when reservoir levels are high. In some cases, this temporary storage of surface water in the ground-water system has been found to be a significant factor in reservoir management (see Box Q).

Human-controlled reservoir releases and accumulation of water in storage may cause high flows and low flows to differ considerably in magnitude and timing compared to natural flows. As a result, the environmental conditions in river valleys downstream from a dam may be altered as organisms try to adjust to the modified flow conditions. For example, the movement of water to and from bank storage under controlled conditions would probably be much more regular in timing and magnitude compared to the highly variable natural flow conditions, which probably would lead to less biodiversity in river systems downstream from reservoirs. The few studies that have been made of riverine ecosystems downstream from a reservoir indicate that they are different from the pre-reservoir conditions, but much more needs to be understood about the effects of reservoirs on stream channels and riverine ecosystems downstream from dams.

REMOVAL OF NATURAL VEGETATION

To make land available for agriculture and urban growth, development sometimes involves cutting of forests and removal of riparian vegetation and wetlands. Forests have a significant role in the hydrologic regime of watersheds. Deforestation tends to decrease evapotranspiration, increase storm runoff and soil erosion, and decrease infiltration to ground water and base flow of streams. From the viewpoint of water-resource quality and management, the increase in storm runoff and soil erosion and the decrease in base flow of streams are generally viewed as undesirable.

In the western United States, removal of riparian vegetation has long been thought to result in an increase in streamflow. It commonly is believed that the phreatophytes in alluvial valleys transpire ground water that otherwise would flow to the river and be available for use (see Box R). Some of the important functions of riparian vegetation and riparian wetlands include preservation of aquatic habitat, protection of the land from erosion, flood mitigation, and maintenance of water quality. Destruction of riparian vegetation and wetlands removes the benefits of erosion control and flood mitigation, while altering aquatic habitat and chemical processes that maintain water quality.

Effects of Surface-Water Reservoirs on the Interaction of Ground Water and Surface Water

The increase of water levels in reservoirs causes the surface water to move into bank storage. When water levels in reservoirs are decreased, this bank storage will return to the reservoir. Depending on the size of the reservoir and the magnitude of fluctuation of the water level of the reservoir, the amount of water involved in bank storage can be large. A study of bank storage associated with Hungry Horse Reservoir in Montana, which is part of the Columbia River system, indicated that the amount of water that would return to the reservoir from bank storage after water levels are lowered is large enough that it needs to be considered in the reservoir management plan for the Columbia River system. As a specific example, if the water level of the reservoir is raised 100 feet, held at that level for a year, then lowered 100 feet, the water that would drain back to the reservoir during a year would be equivalent to an additional 3 feet over the reservoir surface. (Information from Simons, W.D., and Rorabaugh, M.I., 1971, Hydrology of Hungry Horse Reservoir, northwestern Montana: U.S. Geological Survey Professional Paper 682.)

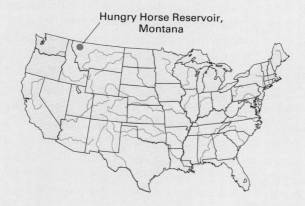

Hungry Horse Reservoir, Montana. (Photograph courtesy of Hungry Horse News.)

Effects of the Removal of Flood-Plain Vegetation on the Interaction of Ground Water and Surface Water

In low-lying areas where the water table is close to land surface, such as in flood plains, transpiration directly from ground water can reduce ground-water discharge to surface water and can even cause surface water to recharge ground water (see Figure 7). This process has attracted particular attention in arid areas, where transpiration by phreatophytes on flood plains of western rivers can have a significant effect on streamflows. To assess this effect, a study was done on transpiration by phreatophytes along a reach of the Gila River upstream from San Carlos Reservoir in Arizona. During the first few years of the 10-year study, the natural hydrologic system was monitored using observation wells, streamflow gages, and meteorological instruments. Following this initial monitoring period, the phreatophytes were removed from the flood plain and the effects on streamflow were evaluated. The average effect of vegetation removal over the entire study reach was that the Gila River changed from a continually losing river for most years before clearing to a gaining stream during some months for most years following clearing. Specifically, average monthly values of gain or loss from the stream indicated that before clearing, the river lost water to ground water during all months for most years. After clearing, the river gained ground-water inflow during March through June and during September for most years (Figure R–1).

Gila River, Arizona

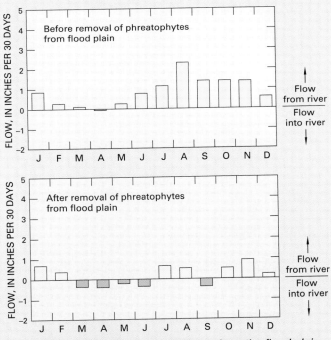

Figure R–1. Removal of phreatophytes from the flood plain along a losing reach of the Gila River in Arizona resulted in the river receiving ground-water inflow during some months of the year. (Modified from Culler, R.C., Hanson, R.L., Myrick, R.M., Turner, R.M., and Kipple, F.P., 1982, Evapotranspiration before and after clearing phreatophytes, Gila River flood plain, Graham County, Arizona: U.S. Geological Professional Paper 655–P.)

Gila River, Arizona. (Photograph by Gregory Pope.)

Modifications to the Atmosphere

ATMOSPHERIC DEPOSITION

Atmospheric deposition of chemicals, such as sulfate and nitrate, can cause some surface-water bodies to become acidic. Concern about the effects of acidic precipitation on aquatic ecosystems has led to research on the interaction of ground water and surface water, especially in small headwaters catchments. It was clear when the problem was first recognized that surface-water bodies in some environments were highly susceptible to acidic precipitation, whereas in other environments they were not. Research revealed that the interaction of ground water and surface water is important to determining the susceptibility of a surface-water body to acidic precipitation (see Box S). For example, if a surface-water body received a significant inflow of ground water, chemical exchange while the water passed through the subsurface commonly neutralized the acidic water, which can reduce the acidity of the surface water to tolerable levels for aquatic organisms. Conversely, if runoff of acidic precipitation was rapid and involved very little flow through the ground-water system, the surface-water body was highly vulnerable and could become devoid of most aquatic life.

"The interaction of ground water and surface water is important to determining the susceptibility of a surface-water body to acidic precipitation"

GLOBAL WARMING

The concentration of gases, such as carbon dioxide (CO_2) and methane, in the atmosphere has a significant effect on the heat budget of the Earth's surface and the lower atmosphere. The increase in concentration of CO_2 in the atmosphere of about 25 percent since the late 1700s generally is thought to be caused by the increase in burning of fossil fuels. At present, the analysis and prediction of "global warming" and its possible effects on the hydrologic cycle can be described only with great uncertainty. Although the physical behavior of CO_2 and other greenhouse gases is well understood, climate systems are exceedingly complex, and long-term changes in climate are embedded in the natural variability of the present global climate regime.

Surficial aquifers, which supply much of the streamflow nationwide and which contribute flow to lakes, wetlands, and estuaries, are the aquifers most sensitive to seasonal and longer term climatic variation. As a result, the interaction of ground water and surface water also will be sensitive to variability of climate or to changes in climate. However, little attention has been directed at determining the effects of climate change on shallow aquifers and their interaction with surface water, or on planning how this combined resource will be managed if climate changes significantly.

Effects of Atmospheric Deposition on the Quality of Ground Water and Surface Water

In areas where soils have little capacity to buffer acids in water, acidic precipitation can be a problem because the infiltrating acidic water can increase the solubility of metals, which results in the flushing of high concentrations of dissolved metals into surface water. Increased concentrations of naturally occurring metals such as aluminum may be toxic to aquatic organisms. Studies of watersheds have indicated that the length of subsurface flow paths has an effect on the degree to which acidic water is buffered by flow through the subsurface. For example, studies of watersheds in England have indicated that acidity was higher in streams during storms when more of the subsurface flow moved through the soil rather than through the deeper flow paths (Figure S–1). Moreover, in a study of the effects of acid precipitation on lakes in the Adirondack Mountains of New York, the length of time that water was in contact with deep subsurface materials was the most important factor affecting acidity because contact time determined the amount of buffering that could take place (Figure S–2).

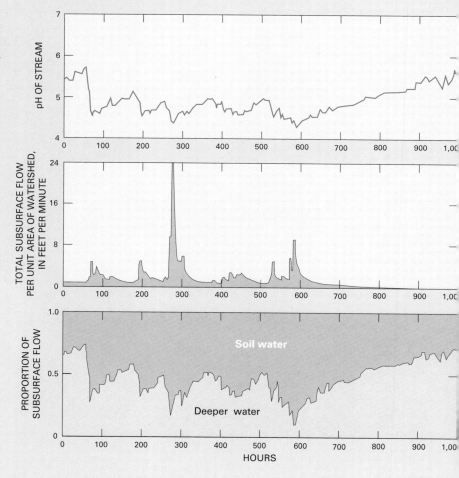

Figure S–1. Acidity is higher (pH is lower) in streams when most of the flow is contributed by shallow soil water because the water has had less time to be neutralized by contact with minerals compared to water that has traversed deeper flow paths. (Modified from Robson, A., Beven, K.J., and Neal, C., 1992, Towards identifying sources of subsurface flow— A comparison of components identified by a physically based runoff model and those determined by chemical mixing techniques: Hydrological Processes, v. 6, p. 199–214.) (Reprinted with permission from John Wiley & Sons Limited.)

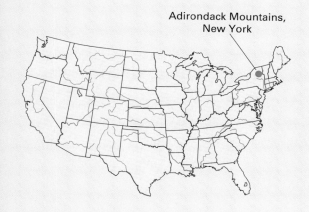

Panther Lake in Adirondack Mountains, New York. (Photograph by Douglas Burns.)

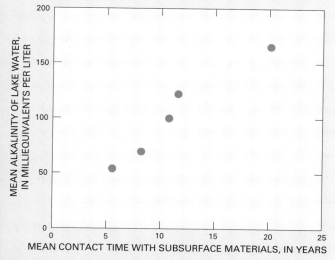

Figure S–2. *The longer water is in contact with deep subsurface materials in a watershed, the higher the alkalinity in lakes receiving that water. (Modified from Wolock, D.M., Hornberger, G.M., Beven, K.J., and Campbell, W.G., 1989, The relationship of catchment topography and soil hydraulic characteristics to lake alkalinity in the northeastern United States: Water Resources Research, v. 25, p. 829–837.)*

CHALLENGES AND OPPORTUNITIES

The interaction of ground water and surface water involves many physical, chemical, and biological processes that take place in a variety of physiographic and climatic settings. For many decades, studies of the interaction of ground water and surface water were directed primarily at large alluvial stream and aquifer systems. Interest in the relation of ground water to surface water has increased in recent years as a result of widespread concerns related to water supply; contamination of ground water, lakes, and streams by toxic substances (commonly where not expected); acidification of surface waters caused by atmospheric deposition of sulfate and nitrate; eutrophication of lakes; loss of wetlands due to development; and other changes in aquatic environments. As a result, studies of the interaction of ground water and surface water have expanded to include many other settings, including headwater streams, lakes, wetlands, and coastal areas.

Issues related to water management and water policy were presented at the beginning of this report. The following sections address the need for greater understanding of the interaction of ground water and surface water with respect to the three issues of water supply, water quality, and characteristics of aquatic environments.

Water Supply

Water commonly is not present at the locations and times where and when it is most needed. As a result, engineering works of all sizes have been constructed to distribute water from places of abundance to places of need. Regardless of the scale of the water-supply system, development of either ground water or surface water can eventually affect the other. For example, whether the source of irrigation water is ground water or surface water, return flows from irrigated fields will eventually reach surface water either through ditches or through ground-water discharge. Building dams to store surface water or diverting water from a stream changes the hydraulic connection and the hydraulic gradient between that body of surface water and the adjacent ground water, which in turn results in gains or losses of ground water. In some landscapes, development of ground water at even a great distance from surface water can reduce the amount of ground-water inflow to surface water or cause surface water to recharge ground water.

The hydrologic system is complex, from the climate system that drives it, to the earth materials that the water flows across and through, to the modifications of the system by human activities. Much research and engineering has been devoted to the development of water resources for water supply. However, most past work has concentrated on either surface water or ground water without much concern about their interrelations. The need to understand better how development of one water resource affects the other is universal and will surely increase as development intensifies.

Water Quality

For nearly every type of water use, whether municipal, industrial, or agricultural, water has increased concentrations of dissolved constituents or increased temperature following its use. Therefore, the water quality of the water bodies that receive the discharge or return flow are affected by that use. In addition, as the water moves downstream, additional water use can further degrade the water quality. If irrigation return flow, or discharge from a municipal or industrial plant, moves downstream and is drawn back into an aquifer because of ground-water withdrawals, the ground-water system also will be affected by the quality of that surface water.

Application of irrigation water to cropland can result in the return flow having poorer quality because evapotranspiration by plants removes some water but not the dissolved salts. As a result, the dissolved salts can precipitate as solids, increasing the salinity of the soils. Additional application of water dissolves these salts and moves them farther downgradient in the hydrologic system. In addition, application of fertilizers and pesticides to cropland can result in poor-quality return flows to both ground water and surface water. The transport and fate of contaminants caused by agricultural practices and municipal and industrial discharges are a widespread concern that can be addressed most effectively if ground water and surface water are managed as a single resource.

Water scientists and water managers need to design data-collection programs that examine the effects of biogeochemical processes on water quality at the interface between surface water and near-surface sediments. These processes can have a profound effect on the chemistry of ground water recharging surface water and on the chemistry of surface water recharging ground water. Repeated exchange of water between surface water and near-surface sediments can further enhance the importance of these processes. Research on the interface between ground water and surface water has increased in recent years, but only a few stream environments have been studied, and the transfer value of the research results is limited and uncertain.

The tendency for chemical contaminants to move between ground water and surface water is a key consideration in managing water resources. With an increasing emphasis on watersheds as a focus for managing water quality, coordination between watershed-management and ground-water-protection programs will be essential to protect the quality of drinking water. Furthermore, ground-water and surface-water interactions have a major role in affecting chemical and biological processes in lakes, wetlands, and streams, which in turn affect water quality throughout the hydrologic system. Improved scientific understanding of the interconnections between hydrological and biogeochemical processes will be needed to remediate contaminated sites, to evaluate applications for waste-discharge permits, and to protect or restore biological resources.

Characteristics of Aquatic Environments

The interface between ground water and surface water is an areally restricted, but particularly sensitive and critical niche in the total environment. At this interface, ground water that has been affected by environmental conditions on the terrestrial landscape interacts with surface water that has been affected by environmental conditions upstream. Furthermore, the chemical reactions that take place where chemically distinct surface water meets chemically distinct ground water in the hyporheic zone may result in a biogeochemical environment that in some cases could be used as an indicator of changes in either terrestrial or aquatic ecosystems. The ability to understand this interface is challenging because it requires the focusing of many different scientific and technical disciplines at the same, areally restricted locality. The benefit of this approach to studying the interface of ground water and surface water could be the identification of useful biological or chemical indicators of adverse or positive changes in larger terrestrial and aquatic ecosystems.

Wetlands are a type of aquatic environment present in most landscapes; yet, in many areas, their perceived value is controversial. The principal characteristics and functions of wetlands are determined by the water and chemical balances that maintain them. These factors in large part determine the value of a wetland for flood control, nutrient retention, and wildlife habitat. As a result, they are especially sensitive to changing hydrological conditions. When the hydrological and chemical balances of a wetland change, the wetland can take on a completely different function, or it may be destroyed. Generally, the most devastating impacts on wetlands result from changes in land use. Wetlands commonly are drained to make land available for agricultural use or filled to make land available for urban and industrial development. Without understanding how wetlands interact with ground water, many plans to use land formerly occupied by wetlands fail. For example, it is operationally straightforward to fill in or drain a wetland, but the ground-water flow system that maintains many wetlands may continue to discharge at that location. Many structures and roads built on former wetlands and many wetland restoration or construction programs fail for this reason. Saline soils in many parts of the central prairies also result from evaporation of ground water that continues to discharge to the land surface after the wetlands were drained.

Riparian zones also are particularly sensitive to changes in the availability and quality of ground water and surface water because these ecosystems commonly are dependent on both sources of water. If either water source changes, riparian zones may be altered, changing their ability to provide aquatic habitat, mitigate floods and erosion, stabilize shorelines, and process chemicals, including contaminants. Effective management of water resources requires an understanding of the role of riparian zones and their dependence on the interaction of ground water and surface water.

ACKNOWLEDGMENTS

Technical review of this Circular was provided by S.P. Garabedian, J.W. LaBaugh, E.M. Thurman, and K.L. Wahl of the U.S. Geological Survey, and James Goeke of the Nebraska Conservation and Survey Division, University of Nebraska. J.V. Flager provided technical and editorial reviews of the manuscript at several stages during its preparation, and M.A. Kidd edited the final manuscript. Design and production of the Circular were led by R.J. Olmstead. Conceptual landscapes were provided by J.M. Evans, and manuscript preparation was provided by J.K. Monson.